OPTICAL COMPUTING

OPTICAL COMPUTING

Proceedings of the Thirty-Fourth Scottish
Universities Summer School in Physics
Heriot-Watt University, Edinburgh, August 1988
A NATO Advanced Study Institute

Edited by
B.S. Wherrett & F.A.P. Tooley
Heriot-Watt University, Edinburgh, Scotland

Published by the
Scottish Universities Summer School in Physics

SUSSP PUBLICATIONS

Edinburgh University Physics Department
King's Buildings, Mayfield Road
Edinburgh EH9 3JZ, Scotland

Marketed and distributed worldwide by
IOP Publishing Ltd (Adam Hilger)
Techno House, Redcliffe Way, Bristol BS1 6NX, England
and
335 East 45th Street, New York, NY 10017-3483, USA

ISBN 0 905945 17 4

Produced by Edinburgh University Press
and printed in Great Britain by
Redwood Burn Ltd, Trowbridge

SUSSP PROCEEDINGS

DIRECTOR'S PREFACE

The 34th Scottish Universities Summer School in Physics was held at the campus of Heriot-Watt University from 14th-26th August 1988. The School was supported by grants from the NATO Scientific Affairs Division programme for Advanced Study Institutes, from the Scottish Universities, and from the Science & Engineering Research Council of the United Kingdom.

Close to one hundred participants, from over twenty countries, attended the School. Interactions between students, lecturers and organisers were lively and informative, both during the lecture programme and throughout the social activities. The subject of optical computing is truly interdisciplinary, attendees had backgrounds including optical and solid-state physics, electronic engineering, computer science and mathematics. The lecturers are to be thanked for their skillful presentations, that afforded tutorial level information to those unfamiliar with the breadth of the subject and sufficient state-of-art details and overviews to maintain the interest of all.

I am indepted to the staff of Heriot-Watt University for the smooth running of all aspects of the School, and in particular to the many members of the Department of Physics whose unflagging help produced a Summer School atmosphere which will be long remembered by all who attended.

B.S. Wherrett

ORGANIZING COMMITTEE

Prof. B.S. Wherrett	Director
Dr. F.A.P. Tooley	Secretary
Dr. H.A. MacKenzie	Treasurer
Mr. J. Smith	Steward

(Department of Physics, Heriot-Watt University, Scotland, UK)

INTERNATIONAL ADVISORY COMMITTEE

Dr. J.P Huignard	Thomson CSF, Orsay	France
Prof. A.W. Lohmann	Erlangen University	West Germany
Dr. D.A.B. Miller	AT&T Bell Labs., Holmdel	USA
Prof. S.D. Smith	Heriot-Watt University	UK
Prof. G.S. Pawley	Edinburgh University	UK

LECTURERS

Dr. K.H. Bohle	IBM, Stuttgart	West Germany
Dr. K.-H. Brenner	Erlangen University	West Germany
Prof. S.A. Collins	Ohio State University	USA
Prof. J.W. Goodman	Stanford University	USA
Prof. A.W. Lohmann	Erlangen University	West Germany
Dr. D.A.B. Miller	AT&T Bell Labs., Holmdel	USA
Prof. G. Parry	University College London	UK
Prof. D. Psaltis	California Institute of Technology	USA
Dr. H. Rajbenbach	Thomson CSF, Orsay	France
Prof. S.D. Smith	Heriot-Watt University	UK
Prof. D.J. Wallace	Edinburgh University	UK

LIST OF PARTICIPANTS

Dr. G.R. Allan, CNRS-SNCI, France

Mr. D. Andonov, Warwick Univ., U.K.

Mr. U. Becker, Univ. Kaiserslautern, FRG

Mr. D. Berard, Univ. Paris-Sud, France

Dr. L.M. Bernardo, Univ. Portugal, Portugal

Mr. A.J. Bostel, Kings College London, U.K.

Dr. L.E.M. Brackenbury, Univ. Manchester, U.K.

Mr. D. Brady, California Inst. of Tech., U.S.A.

Mr. G. Buller, Heriot-Watt Univ., U.K.

Dr. J.-A. Cavailles, LEPA, France

Mr. F. Chataux, IOTA, France

Mr. M. Claydon-Smith, Lancaster Univ., U.K.

Mr. A.R. Corless, Thorn EMI, U.K.

Mr. R. Craig, Heriot-Watt Univ., U.K.

Mr. W.A. Crossland, STC Technology Ltd., U.K.

Mr. R. Damm, Fried. Krupp GmbH, FRG

Dr. M. Derstine, Boeing Electronics, U.S.A.

Mr. C. Dornfeld, Odense Univ., Denmark

Dr. W. Dultze, Deutsche Bundespost, FRG

Dr. R. Ellialtioglu, Bilkent Univ., Turkey

Mr. V.C. Esch, OSC, Univ. of Arizona, U.S.A.

Prof. J. Ferreira da Rocha, Univ. of Aveiro, Portugal

Mrs. S. Fisher, Heriot-Watt Univ., U.K.

Ms. E. Giorgetti, IROE-CMR, Italy

Mr. A.P. Gracian, Univ. College London, U.K.

Dr. D.J. Hagan, CREOL, U.S.A.

Mr. C. Haile, OCLI, U.K.

Dr. A. Hartmann, MCC-ACA/STL, U.S.A.

Mr. F. Hrebabetsky, Univ. Munchen, FRG

Prof. Y. Ichioka, Osaka Univ., Japan

Dr. V. Ignatov, USSR Academy of Science, U.S.S.R.

Dr. Y. Iyechika, Univ. Munster, FRG

Dr. Y.D. Kalafati, USSR Academy of Science, U.S.S.R.

Dr. P.W. King, Paisley College of Tech., U.K.

Mr. J.M. Kinser, Univ. of Alabama, U.S.A.

Mr. B. Kippelen, IPCMS, France

Prof. C. Klingshirn, Univ. Kaiserslautern, FRG

Mr. M. Kunz, Univ. Kaiserslautern, FRG

Mr. T.-Y.D. Lam, Cambridge Univ., U.K.

Mr. A.R. MacGregor, Univ. of Edinburgh, U.K.

Mr. P.W. McOwan, Kings College London, U.K.

Mr. C. Marriott, LROL, Canada

Ms. K.B. Mason, Manchester Univ., U.K.

Dr. G. Mendes, Imperial College, U.K.

Mr. T. Mikropoulos, Theoret. & Phys. Chem. Inst., Greece

Dr. P.K. Milsom, RSRE, U.K.

Mr. I. Muirhead, OCLI, U.K.

Mr. D. Nabors, Stanford Univ., U.S.A.

Mr. J. Oberle, Univ. Strasbourg, France

Mr. U. Olin, Royal Inst. Tech., Sweden

Mr. A. Oral, Bilkent Univ., Turkey

Mr. G. Pratesi, IROE-CNR, Italy

Dr. S. Redfield, MCC-ACA/STL, U.S.A.

Mr. P. Refregier, Thomson CSF, France

Mr. B. Robertson, Heriot-Watt Univ., U.K.

Mr. J. Schwider, Univ. Erlangen-Nurnberg, FRG

Mr. D. Selviah, Univ. College London, U.K.

Dr. P.K. Sen, Bhopal Univ., India

Mr. M. Shabeer, Univ. of Strathclyde, U.K.

Dr. M. Sidrach de Cardona, CIEMAT, Spain

Dr. E.W. Smith, Massey Univ., New Zealand

Mr. J. Snowdon, Heriot-Watt Univ., U.K.

Mr. K. Stephen, Univ. of Strathclyde, U.K.

Mr. C. Stirk, California Inst. of Tech., U.S.A.

Mr. M. Taylor, Univ. College London, U.K.

Dr. P. Urquhart, BTRL, U.K.

Dr. J. Traff, Tech. Univ. Denmark, Denmark

Dr. E.Y. Tsiang, MBI Microbeam Inc., U.S.A.

Dr. T. Van Eck, Univ. California, U.S.A.

Mr. C. Van Hoof, IMEC, Belgium

Mr. N.A. Vainos, Univ. of Essex, U.K.

Mr. A.H.A. Vasara, Helsinki Univ. of Tech., Finland

Dr. D. Vass, Univ. of Edinburgh, U.K.

Mr. J.M. Wang, Univ. de Paris-Sud, France

Mr. K. Weible, Univ. de Neuchatelle, Switzerland

Dr. K. Welford, RSRE, U.K.

Mr. T. Wicht, Johann-Wolfgang Goethe Univ., FRG

Dr. D. Williams, Marconi Research Centre, U.K.

Mr. A. Witt, Univ. Kaiserslautern, FRG

Prof. T. Yatagai, Univ. of Tsukuba, Japan

PHOTOGRAPH

Row 1 (front)

S. Gilroy, A.C. Walker, M.R. Taghizadeh, J. Smith, J. McClelland, F.A.P. Tooley, H.A. MacKenzie, K.-H. Bohle, S.D. Smith, S. Redfield, B.S. Wherrett, A. Lohmann, S.A. Collins Jr., H. Rajbenbach, D.A.B. Miller, J.W. Goodman, A.K. Kar, L.E.M. Brackenbury

Row 2

W. Dulz, N.A. Vainos, F. Chataux, A. Rashed, Yu.D. Kalafati, C. Stirk, P. Milsom, W. Ji, J. Schwider, A.K. Iltaif, T. Yatagai, D. Nabors, D.J. Hagan, K.J Weible, A.D. Lloyd, A. Oral, C.F. Klingshirn, M. Sidrach de Cardona, R. Damm, A. Gracian, D. Selviah, E.W. Smith, V. Esch, J. Kinser, I.R. Agool, E.Y. Tsiang

Row 3

M. Kunz, T. Mikropoulos, A. Witt, C. Van Hoof, U. Becker, W. Crossland, C. Marriott, J. Snowdon, D. Berard, L. Bernardo, M. Shabeer, R. Ellialtioglu, S.P. Fisher, K. Mason, M. Claydon-Smith, J. Ferriera da Rocha, P.W. King, D. Williams, T.Y.D. Lam, J.M. Wang, E. Giorgetti, F. Hrebabetzky, G. Pratesi, P.K. Sen, G.F. Mendes

Row 4

T. Wicht, Y. Iyechika, C. Dornfeld, K. Welford, M. Derstine, D. Brady, V. Ignatov, A. Cavailles, A. Corless, I.T. Muirhead, A. MacGregor, C. Haile, T.E. Van Eck, U. Olin, M. Taylor, A. Vasara, J. Traff, P. Refregier, J. Oberle, B. Kippelen, P. Urquhart, K. Stephen, A. Darzi, A. Hartmann

CONTENTS

INTRODUCTION

Brian S. Wherrett

Department of Physics

Heriot-Watt University

Edinburgh, Scotland, UK

The uses and prospects for optics in information processing are now wide ranging; from spatial filtering of images by attenuating specific Fourier components through to proposals for general purpose optical computers. A definition and overview of the subject is given by J.W. Goodman in Chapter 2 of this edited text. The purpose of this introduction is to set the various contributions of these Summer School Proceedings in context.

Optical computing, as a research field, can be thought of as starting in 1964 with the first publication of a collection of papers, presented at the Symposium on Optical and Electrooptical Information Processing [1]. It is no coincidence that this meeting was held just four years after the invention of the laser, with the dramatic increase in information processing ability afforded by coherent, intense, narrow bandwidth radiation. Since that time one can point to at least 19 conference proceedings and journal special issues devoted to research publications in the field [2].

It is less than easy to draw from this collation of papers the threads of the subject. Individual topics appear over a few meetings and then disappear from the research literature, often without apparent influence on other topics. The absence until recently of a textbook covering the subject as a whole compounds the confusion of researchers entering the field [3]. It is therefore hoped that this tutorial level text will help establish the structure and bounds of the subject.

A useful division of the field of optical information processing is into:
(i) Techniques, (ii) Algorithms, (iii) Architectures and (iv) Applications. This division immediately emphasises the interdisciplinary nature of the subject. *Techniques* are traditionally the realm of physicists and electronics engineers, involving optical phenomena, the interaction mechanisms of light and matter, and the construction of devices that exploit these mechanisms in a controllable and efficient manner. The fabrication processes and the materials advancement involve also the expertise of materials scientists and chemists. *Algorithm* development is essentially a mathematics topic, it covers the representation of numeric or physical data and the arithmetic or logic-based methods by which data can be manipulated to solve a given problem. The computer scientist is concerned with the *Architecture* by which the components developed under the techniques heading are put together in a system that efficiently implements the algorithms on the data presented to it. Finally the construction of any processing system must be aimed towards one or more *Applications*. Therefore the existing and perceived needs of research, consumer or military markets must be borne in mind, also the achievements and potential of alternative information processing systems.

A textbook on optical computing could usefully be subdivided as described above. An attempt has been made to structure this proceedings along these lines. It must be borne in mind, however, that the invited experts are each involved with more than one aspect of the subject, often being concerned both with the fundamental science and optimisation of particular devices and with the use of these devices in optical processing circuitry. I believe that it is fair to say however that the research described in Chapters 3 - 10 is predominantly 'techniques-led', that of Chapters 11 - 14 concerns mainly algorithms and architectures, whilst the final two chapters present the relevant achievements of electronic computing. Applications of optical computing are addressed variously throughout.

Any information processing system requires (i) optical sources, (ii) a method for modulating the sources in order to carry the information, (ii) an information processing stage in which decisions are made and the information may be altered, (iv) optical communication & interconnection, (v) detection and (vi) display and/or storage. A techniques section of a definitive optical computing text could contain chapters on each of these components. There are of course entire texts devoted to laser sources, optical fibre communication, detectors, displays and optical storage [4]. These topics are still subject to advancement but are not at the early stages of development that properly form the subject of a Summer School Proceedings. We concentrate here on techniques for Information Processing and for Data Interconnection either between the processing elements themselves or between the elements and the input/output stages. Chapters 3 - 7 are concerned primarly, but by no means exclusively, with processing techniques, in Chapters 8 - 10 optical information communication is foremost.

At the fundamental level the striking difference between optics and electronics is that there is effectively no interaction between photons travelling in vacuum whereas the Coulombic interaction between electrons is very strong. Consequently optical communication without crosstalk of streams of data and 1-D or 2-D images is easy, whereas electronic communications are restricted in zero or 1-D and impossible in highly parallel 2-D format. It is the combination of massive parallelism and cross-talk and interference free interconnection that gives optics its biggest advantage over electronics. These Proceedings concentrate on areas where this advantage is likely to have its greatest potential, namely in the manipulation of two-dimensional images.

Given the absence of direct photon-photon interactions it is essential that indirect interactions are employed if optical image processing is to be achieved, it is then a matter of the degree to which material excitation (charge displacement) is involved. One option is to detect the light, the electronic signals are processed electronically, and the resulting information interfaced back to optics via a modulator. In this way the interconnect freedom of optics can be used in combination with the processing power of electronic circuitry; limitations, at present lie with the interfacing input and output. Viewed as a black box the above device is simply a transmitter (or reflector) with a nonlinear optical transfer function, the function itself being controlled by the address to the electronic circuitry. There are several alternative methods to catalyse the interaction of photons with photons.

In Chapter 3 optically addressed spatial light modulators are described; here for example an optical field incident on one medium generates a static electric field that in turn is used to alter the refractive index of a second medium and hence the transfer properties of the electro-optic medium to a second optical field. In Chapters 4 and 7, bistable devices that rely on either the photo-generation of carriers in semiconductors or the subsequent material heating, followed by a refractive index change (of the same material), are described. Similar devices in which the absorption rather than the refraction dominates the transfer nonlinearity are discussed in Chapter 5; static electric fields induced optically are used to produce relatively large absorption changes - the self-electroabsorptive effect. None of the above devices relies on coherent input light. Therefore, if sensitive enough they can be used as incoherent-to-coherent converters. This property is exploited in the use of spatial light modulators in Fourier optics, Chapter 3. The photorefractive devices described in Chapter 7 also rely on the generation of static fields. In this case the radiation absorption generates carriers that diffuse to neighbouring trap sites. The resulting space-charge fields create refractive index changes via the electrooptic effect and hence again produce a nonlinear transfer.

In the 1980's one of the most active areas of optical computing has been the development of the above techniques for the optical control of optical information and of the exploitation of relatively crude devices in prototype optical processing circuits. Such circuits are described in Chapter 6. In each case the processing device is an optically addressed 2-D spatial light modulator. Other components in processing circuits are of course also 2-D light modulators, but with quite different modes of address. Thus a single lens is a 2-D phase modulator, the address being the mechanically controlled fabrication of the thickness variation. A hologram has structurally controlled index modulation, a computer generated hologram has a controlled attenuation modulation. These fixed, rather than real-time variable, devices are key to many optical interconnection schemes; example uses are presented in Chapter 8.

Real-time modulators that are used for the transfer of information between one discipline and another, or from one dimensionality to another, include acousto-optic Bragg cell devices, electrically addressed 2-D spatial light modulators - index changes brought about by lattice vibrations or electric fields are again the mechanism by which the light-matter interaction is controlled. The former are mentioned in Chapter 2; the techniques of acousto-optic information processing are now well established (although not widely applied), details can be found in a number of monograph references [5]. Externally addressed 2-D light modulators are discussed in Chapter 3.

From the optics viewpoint the algorithms of data processing can be categorised by the amount of fan-out demanded of each signal. Thus in an electronic processing scheme it might be necessary only to achieve optical Data Reordering, the output from one device being fed to only one detector. The 'perfect shuffle' for example is the basis for fast sorting algorithms and for the fast discrete Fourier transform. Clock-distribution (Chapter 10), however, would demand one-to-many fan-out, with likely restrictions on uniformity and registration of the generated beam arrays.

For processing itself the output of any one active device must usually be fed to two or more further devices, and equally each device must be able to accept at least two information signals. If the fan-in/out is small, less than 1-to-10 say, one is usually concerned with Digital Optics. Here the values of 'zero' and 'one' levels representing binary numeric or image data are to be monitored accurately throughout calculations. The number representation, algorithm and architectures for parallel processing of digital optics are presented particularly in Chapters 11 - 13. There is presently a clear division between proposed cellular 2-D schemes for digital optics processing and schemes with multiple fan-out/in presently proposed for optical associative memory and neural networks. The latter are discussed in Chapter 14. With fanning of perhaps 1-to-100 up to 1-to-10^4 it is not possible to maintain the accuracy of digital arithmetic; one is concerned with decision making by thresholding the combined input to a given processor at some approximate level. I term this approach Threshold Optics. Finally the global fan-in/out achieved by lens and diffraction optics, is the basis for the established field of Fourier Optics [6] and is represented herein in the use of photorefractive devices for real-time holography and reconfigurable optical interconnects (Chapters 7 and 10).

The final two Chapters (15 and 16) are included in order to place the achievements of optics in perspective and to indicate those areas of 'computing' in which optics is likely to best complement (rather than supercede) electronic machines. These Chapters cover briefly the hardware developments of electronics, parallel architectures, and the use of parallel architectures for specific applications including neural networks.

There is no way that the Proceedings of a single, two-week Summer School could hope to encompass the full range of topics under the general heading of Optical Computing. It is hoped that those topics selected will give the readers a feel for the subject as a whole however as well as details of some of the most active areas of development that are currently being pursued.

References

1. Proc. Symp. on Optical & Electrooptical Information Processing, Boston 1964, Ed. J. Tippett et al, MIT Press 1968.

2. cf. the following Conference Proceedings & Journal Special Issues. IEEE Trans. Comp. C-24, April 1975. Special Issue on Optical Computing.
   Proc. IEEE 65, Jan. 1977. Special Issue on Optical Computing.
   SPIE 231/2, Proc. Int. Optical Computing Conf., April 1980.
   SPIE 388, Adv. in Optical Information Processing, Jan. 1983.
   SPIE 422, Proc. 10th Int. Optical Computing Conf., April 1983.
   Opt. Eng., 23 Jan/Feb 1984. Special Issue on Optical Computing.
   SPIE 456, Optical Computing, Jan. 1984.
   Proc. IEEE 72, July 1984. Special Issue on Optical Computing.

Technical Digest of the 13th congress of the International Commission for Optics, Sapporo, Aug. 1984.

Opt. Eng. 24, Jan/Feb 1985. Special Issue on Optical Computing & Optical Information Processing Components.

Technical Digest of OSA topical meeting on Optical Computing, Incline Village, Nevada, March 1985.

Opt. Eng. 25, Jan. 1986. Special Issue On Digital Optical Computing.

SPIE 625, Optical Computing, Jan. 1986

SPIE 634, Institute for Advanced Optical Technologies, Optical & Hybrid Computing, March 1986.

Applied Optics 25, May 1986. Special Issue on Optical Computing.

Opt. Eng. 26, Jan. 1987, Special Issue on Optical Computing & Nonlinear Optical Signal Processing.

Technical Digest of OSA topical meeting on Optical Computing, Incline Village, Nevada, March 1987.

Opt. Eng. 26, March 1987. Special Issue on Optical Information Processing.

ICO Meeting on Optical Computing, Toulon, Aug. 1988, to be published as SPIE 963.

3.   "Optical Computing, A survey for Computer Scientists", D. G. Feitelson, MIT Press 1988.

4.   Example text in areas relevant to Optical Computing.

Lasers: physics, systems & techniques. Proc. 23rd Scottish Universities Summer School in Physics. Ed. W.J. Firth & R.G. Harrison, SUSSP Edinburgh, 1983.

Optical Fibre Communications. T. Senior, Prentice-Hall International Ltd., London, 1985.

Detection of Optical & Infrared Radiation. R.H. Kingston
Springer Series in Optical Sciences. Springer-Verlag NY, 1978.

Principles of Optical Disc Systems. Ed. G. Bouwhuis et al. Adam Hilger Ltd., Bristol UK, 1985.

5.   Acousto-optic Signal Processing. Ed. N.J. Berg & J.N. Lee, Marcel Dekker Inc. NY. Proc. IEEE 69, Jan. 1981. Special Issue on Acousto-Optic Processing.

6.   Introduction to Fourier Optics. J.W. Goodman. McGraw-Hill, NY, 1968.

Fourier Optics: an Introduction. E.G. Steward. John Wiley & Sons, Chichester, UK, 1987.

Optical Data Processing. Springer Topics in Applied Physics 23, Ed. D. Casesant, Springer-Verlag, NY, 1978.

Joseph W. Goodman
Department of Electrical Engineering
Stanford University
Stanford, California 94305 U.S.A.

Background

There is no common agreement as to what the field of "optical computing" actually embraces. Here we adopt the definition that optical computing consists of all methods for intentionally manipulating optically represented data for useful purposes. Admittedly, this is a very broad definition. Most importantly, it includes not only the current attempts to realize digital computers based on optical logic gates, but also the older field of analog optical information processing. This chapter presents a short history of the field, defined in this way, and attempts to show how different sub-fields grew from their predecessors.

Figure 1 illustrates what could be called the "family tree" of optical computing. Each major branch represents a different sub-field. Branches divide into sub-branches, and in some cases the sub-branches become intertwined. So too with the sub-fields of optical computing. The roots of the tree can be said to be the work of Ernst Abbe [1], who first proposed that a coherently illuminated object could be regarded to be a collection of grating components, each of which generates focused diffraction orders in the plane where the illumination source is imaged. Manipulation of these orders changes the grating content of the object, and can markedly modify the character of its image. The upper roots of the tree are the work of Fritz Zernike [2], who recognized that proper manipulation of the grating orders can render a phase object visible without staining, a discovery (the phase-contrast microscope) for which he received a Nobel prize.

During the 1950's, there were few scientists working in the field. Nonetheless, their work had major impact, and forms the lower trunk of the tree. We mention explicitly the work of Marechal [3], Tsujiuchi [4], O'Neill [5], and Lohmann [6].

In 1960 a classic paper by Cutrona, Leith, Palermo and Porcello [7] exposed in the open literature for the first time new ideas previously developed at the University of Michigan regarding applications of coherent optical systems to the processing of radar data. This work was to have enormous impact, lasting even to today, and is shown to compose the upper trunk of the tree.

In the early 1960's, the tree began to fork into branches. As depicted in Fig. 1, the branches are arranged chronologically, with the oldest on the left, and the youngest on the right. Thus time progresses in a clockwise manner.

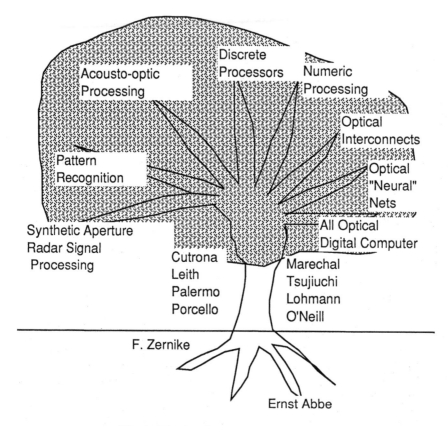

Fig. 1. The family tree of optical computing

Synthetic-Aperture Radar

The oldest of these branches is the one labeled "synthetic aperture radar". The origin of this branch again lay in the work at the University of Michigan, in which coherent optical systems were used to create a kind of scaled analog of a radar system, from which maps of the microwave reflectivity of terrain could be constructed. A radar with a highly stable oscillator is carried by an aircraft or satellite along a straight-line path; along the path, short pulses of energy are emitted, and returning echoes are recorded in both amplitude and phase. Range resolution (normal to the flight path) is obtained by conventional pulse-echo timing. Azimuth resolution (along the flight path) is obtained by compression of the doppler histories of the received signals, treating them as if they were received on a coherent array of antennas distributed along the flight path.

We will not dwell on the operation of this type of radar system here. Of more interest is the type of optical system used for forming terrain maps from the recorded signals. Assuming that the returned signals are recorded on film in the aircraft, upon processing those signals, one finds that the location of the desired image in the range direction coincides with the plane of the recording film, whereas the image in the azimuth direction lies in a *tilted* plane that does not coincide with the film. Thus the optical system must be one that can simultaneously image the film plane and a more remote tilted azimuth plane onto a common plane where the terrain image will be found. An early version of such a processor [8], shown in Fig 2., used a conical lens at the film plane to move the azimuth image to infinity, and a cylindrical lens to move the range image to infinity. The two infinitely distant planes are then brought into coincidence by a spherical lens, where an output slit is used to prevent distortion. Film is drawn across the output slit as the film at the input is moved through the processor, yielding the desired image of the terrain.

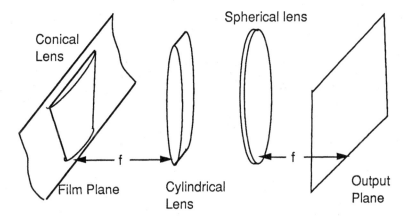

Fig. 2. Early optical processor for synthetic-aperture radar signal processing

An improved optical system, called the "tilted plane processor", was described by Kozma, Leith and Massey in 1972 [9]. In this processor, the input plane and the output plane are both tilted, and an anamorphic telescope is used. The resulting image requires no output slit to prevent distortion. This processor remains today one of the most sophisticated optical computing systems ever built.

Optical Pattern Recognition

A second branch of the tree grew from an intersection of the experience gained in radar signal processing with the emerging needs for automatic pattern recognition. In 1964, Vander Lugt published a classic paper [10] in which he described a new and highly practical method for making coherent optical matched filters. The optical filtering system is the usual one shown in Fig. 3, in which the input data is introduced either by film or by a spatial light modulator. A filter placed in the Fourier (focal) plane modifies the spectrum of the input pattern, following which another lens inverse transforms the spectrum to yield a filtered output.

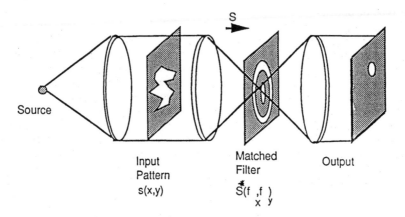

S

Source

Input
Pattern
s(x,y)

Matched
Filter
$\overset{*}{S}(f_x,f_y)$

Output

Fig. 3. Coherent optical matched filter

Vander Lugt's contribution was the demonstration of an interferometric or holographic method for making a so-called "matched" filter that could be used for pattern recognition. The transmittance of such a filter in the frequency domain should be equal to the complex conjugate of the Fourier spectrum of the object one wishes to recognize. The net result will be an output from the processing system consisting of a bright spot at the location of the object within the field of the system, or no bright spot if the object is not present.

The idea of the interferometrically generated matched filter has had enormous intellectual impact. However, to date, it's practical use in real pattern recognition problems has been very limited. Recently a regeneration of interest has resulted from the development of a very compact coherent optical correlator by the Jet propulsion Laboratory and the Perkin Elmer Corporation, aimed at military target recognition [11]. It remains to be seen whether this technology will ever find its way into the commercial marketplace.

A second basic approach to optical pattern recognition is called "diffraction pattern sampling", which bases pattern recognition and classification decisions on distribution of power in the optically obtained Fourier transform of the object [12]. The system is illustrated in Fig. 4. The input is presented to the optical system by means of a spatial light modulator, which converts the incoherent image of the object into a coherent image for processing. The coherent optics Fourier transform the input, displaying its spectrum on a photodetector array. The detector array consists of a series of wedges and rings, which divide the frequency domain into a relatively small number of sub-regions (e.g. 32), within which the incident optical power is measured.

The resulting feature vector is then operated on by an electronic digital system, where any of a number of pattern recognition algorithms can be applied.

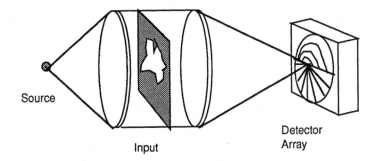

Fig. 4 Diffraction pattern sampling system

This system has the virtue that it reduces the complexity of the power spectrum to a relatively few numbers, which can be readily dealt with by an electronic digital computer. However, this is a true virtue only if the feature vectors obtained in this way form a good basis for recognition of discrimination. A commercial product that uses this approach, together with a simulated neural network for final classification, has recently been reported [13].

More recent work on optical pattern recognition has focused on methods other than the matched filter for classifying patterns. We mention in particular the method known as "synthetic discriminant functions", invented by Braunecker et al [14] and named by Casasent [15].

Acousto-optic Signal Processing

A third major branch of the tree is labeled "acousto-optic signal processing". This branch has a history that dates back to the early 1960, and the work of Rosenthal [16], Slobodin [17], and Arm, Lambert and Weisman [18]. For a review of this area see Ref. [19].

The basis for this approach is the use of a transparent acoustic cell for introducing electrical signals into a coherent optical system. A high-frequency electrical signal, amplitude and phase modulated, is applied to the transducer on an acoustic cell. An acoustic signal is launched into the cell, and induces changes of refractive index by virtue of compression and expansion of the medium. These refractive index changes then spatially phase modulate coherent light passing through the cell in a direction approximately normal to the direction of acoustic propagation.

Early attention was to systems incorporating devices operating in the so-called Raman-Nath regime of acousto-optic diffraction. Such devices are characterized by center frequencies in the few tens of MHz. In recent years attention has shifted to devices operating in the Bragg regime, for which center frequencies can be up to a few GHz. In the latter regime, the cell acts as a "thick" grating, generating only one or two diffraction orders, and offering much greater optical efficiency than Raman-Nath cells.

Two different sub-branches of the acousto-optic branch can be identified, and are called "space integrating" and "time integrating". The most important use of acousto-optic technology to date has been for extremely wide-band space-integrating spectrum analyzers. As illustrated in Fig. 5, the electrical signal to be spectrum analyzed is applied to the transducer on the cell, and the spectrum is formed on a detector array in the focal plane of the Fourier transforming lens. The Fourier transform takes place via an integral over space, hence the term "space integrating". Coherent optical spectrum analyzers with bandwidths as wide as 1 GHz have been made and with time bandwidth products as high as 1000. Integrated optic versions of such systems, using surface acoustic waves, have also been studied and built. A recent review of the field can be found in Ref. [20], and a comprehensive treatment of the subject is found in [21].

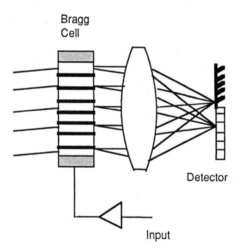

Fig.5. Bragg cell spectrum analyzer

A second architecture in the space-integrating category is the space integrating correlator. A typical configuration is shown in Fig. 6. The incoming RF signal is applied to the transducer of the input cell, and a second reference signal (including a carrier frequency) is stored on a mask. One diffraction order of the transmitted light is imaged onto the mask, and one component of the light diffracted by the mask is focused onto a photodetector. The detected signal is proportional to the squared magnitude of the spatial correlation integral between the incoming signal and the reference, with a shift between the two that depends on time.

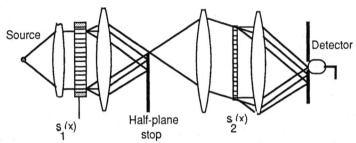

Fig. 6. Space integrating acousto-optic correlator

The sub-branch representing time-integrating correlators began in 1972 with a patent by R.M. Montgomery [22], and was given greater impetus by a publication of Sprague and Koliopolos in 1976 [23]. The architecture is illustrated in Fig. 7. One of the signals to be correlated is input as a traveling acoustic wave in the first acousto-optic cell, while the second signal to be correlated is introduced as a counterpropagating signal in the second cell. The first diffracted orders from each of the cells are selected and caused to interfere on a spatial array of small detectors. Each detector in the array "sees" a different time difference between the two signals, while each detector integrates in time to calculate the correlation for its shift. The correlation integral takes place in time, hence the name "time integrating" correlator.

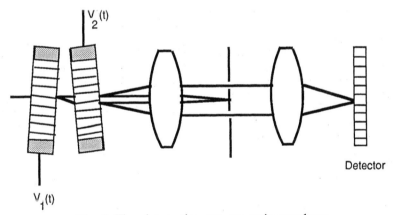

Fig. 7. Time integrating acousto-optic correlator

The significance of the time-integrating architecture lies in the fact that the time bandwidth product achieved no longer is limited by the space-bandwidth product of the optical system (as it is in the space-integrating case). Rather, the practical limitation becomes the buildup of bias terms on the detector. Other forms of time-integrating correlators are known but will not be covered here (see [24], [25]).

Discrete Processors

Discrete processors perform the discrete equivalents of continuous-time linear operations. The general types of operations performed are matrix-vector products and matrix-matrix products. While these systems deal with sampled data, it should be remembered that they are nonetheless analog systems and suffer from all the limitations of accuracy characterizing such systems.

Although there were earlier publications on discrete systems, the field of discrete processing received its initial impetus from the work of Bocker [26], Bromley [27], and Monahan *et al* [28], who proposed an incoherent system for performing matrix-vector multiplications using an LED for a source and a CCD detector. This system, shown in Fig. 8, accepts input data serially, and therefore it would be represented by a sub-branch on our tree labeled "serial". The matrix is stored as a transparency containing NxN cells, where the effective transmittance of each cell is proportional to the value of one corresponding matrix element (assumed non-negative and real). The vector is introduced as a serial set of N current pulses, each with a height proportional to one of the elements of the input vector. The LED then emits a series of light pulses, which pass through the mask and fall on the detector array. The system is best understood by considering first the column of detectors on the left (i.e. furthest into the page). The first current pulse generates a light pulse with strength proportional to the first vector element. This light pulse floods the whole mask, but in particular, a portion of it passes through the left-most column of the mask and deposits change in the left most column of detector elements. Passage through the mask has multiplied this element of the vector by all matrix elements in the left column and has stored these results in the left detector column. Now the charges stored in the detector elements are clocked one column to the right, following which the second LED pulse occurs. The charges in the second detector column now are given a second additive component, namely the product of the second vector element with the matrix elements stored in the second column of the mask. This process repeats, until finally, after N cycles, the last detector column on the right contains the vector that is the product of the stored matrix with the input vector.

Restrictions of non-negativityand realness of the data can be removed by suitable encoding of the bipolar or complex data in terms of non-negative reals [29].

A second sub-branch of the discrete processing branch is called "parallel". In this case, rather than using serial data flow into the processor, we enter the data in parallel, thereby achieving much higher throughput. Figure 9 illustrates the system developed at Stanford in the late 70's [30]. The elements of a vector are introduced in parallel on an array of LED's. The first box labeled "optics" spreads the light in the vertical direction and images in the horizontal direction. Thus the light from each LED becomes a vertical column of light. The matrix mask is the same as that used in the previous system. The second optics box sums the light passed by each row of the

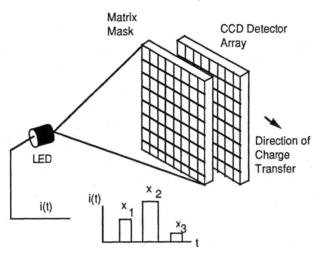

Fig. 8. Serial matrix-vector multiplier

matrix masks and places that light on a separate element of a linear detector array.
Each detected component is then the inner product of the incoming vector with one row
vector of the mask, and the output vector appears on the detector array.

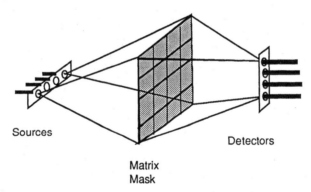

Fig. 9. Parallel matrix-vector multiplier

Such a system is capable of extremely high speed operation by parallelism and the
use of high-speed sources and detectors. If the output is fed back to the input in an
appropriate way, it has been shown that inversion of sets of linear equations can be
performed [31], and eigenvectors can be found [32].   However, these iterative
systems are limited by the effects of noise [33].

The above type of architecture is generally referred to as an "inner product"
architecture, since the basis of operation is a series of vector inner products. Figure
10 shows a "outer product" architecture suggested by Athale [34] for multiplying to

matrices. A row vectors of one matrix is entered on an array of acousto-optic modulators and a column vector of the other matrix is entered on another similar array. Each cycle of the processor generates one outer product of a row vector and a column vector, which is stored on the NxN detector array. The matrix product is built up by summing N such outer products on the time integrating detectors.

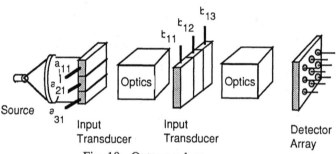

Fig. 10. Outer product processor

The final sub-branch of the discrete processing branch is the one labeled "systolic". This sub-branch is concerned with analog implementations of the systolic architectures originally developed by H.T. Kung [35] and others for VLSI implementation. Figure 11 illustrates the basic building block (part a) and the simplest systolic processor (part b). Referring to part (a) of the figure, an input x representing one element of the input vector arrives from the left. Simultaneously a weighting coefficient a, representing one element of the matrix, arrives from above. The processor, represented by the box, receives input x from the left and passes it to the right unchanged. Simultaneously, it accepts the value y coming from the right and transforms it into the new value y + ax, which continues to move to the left.

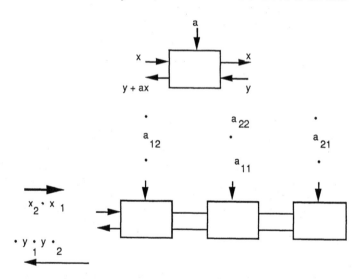

Fig. 11. Systolic processor. (a) Basic building block. (b) Matrix-vector multiplier

Part (b) of the figure illustrates the use of three such units to perform a 2x2 matrix-vector product. The elements of the data vector occupy every other cycle on the upper left input line. The elements of the matrix flow into the system from above, again carefully timed. The product vector flows out of the system to the left.

An analog optical implementation of such a system was first proposed by Caulfield, Rhodes, Foster and Horvitz [36], and various versions of this type of system have been built by Casasent (e.g. see [37]). A fiber-optic systolic processor has been proposed and demonstrated by Tur *et al* [38]. A closely related architecture is the so-called "engagement processor" of Speiser, Whitehouse and independently, Tamura [39].

Numeric Processing

The term "numeric processing" is meant here as an attempt to extract from analog processors greater accuracy than is normally associated with such systems. This is attempted by using the analog system to perform some form of numerical arithmetic.

There are two different approaches to achieving such accuracy that deserve mention here. Each can be regarded as a sub-branch of this branch. The first is residue arithmetic processing [40], and the second is digital multiplication by analog convolution (DMAC) [41].

Residue arithmetic processing utilizes an arithmetic system first discovered in ancient China and having the interesting property that no carries are required in addition, subtraction or multiplication (division is in general not possible in this number system). Each number is represented in terms of a set of residues with respect to several relatively prime bases. The residue is the remainder obtained when the number is divided by the base in question. To add two numbers, we simply add their residues for the first base, add their residues for the second base, etc, and convert the results to residue form. A similar procedure is followed for multiplication. Residues do not interact during the operations, and the residue operations have a cyclic character, suggesting interesting optical implementations. Current efforts are aimed at realizing optical residue processors using very fast table lookup [42].

The DMAC algorithm was first brought to the attention of the optics community by H. Whitehouse, but most significantly was combined with systolic processor ideas by P. Guilfoyle to produce the so-called SAOBIC matrix-vector processor [41], illustrated in Fig. 12. The operation of this processor is rather complex, and the reader is referred to the original reference for details. It suffices to say here that two different Bragg cell arrays are used, one a "slow" cell and the other a "fast" cell. The data entered into this system is in binary form, with each matrix and vector element represented by a series of bits. The input vector is loaded bit-parallel and element-serial, while the elements of the matrix are entered in the form illustrated but bit serial. From the convolutions performed can be found a mixed binary representation of the products of interest. These products must be A/D converted at very high speed and added digitally. The chief practical disadvantage of this approach lies in the large number of very high-performance electronic A/D converters needed.

Remaining Branches

Three main branches remain undiscussed, namely those labeled "optical interconnects", "optical neural nets", and the "all optical digital computer". All of them represent very active areas of research and development, perhaps the most active areas in modern optical computing. These areas will be covered in other chapters in this volume.

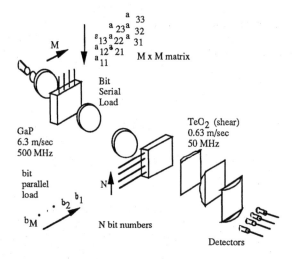

Fig. 12. Saobic Processor

References

1. Ernst Abbe, "Beitrage zur theorie des mikroskopes und der mikroskopischen wahrnehmung", *Archiv. Mikroskopische Anat., 9,* 413-468 (1893).

2. F. Zernike, "Das phasenkontrastverfahren bei der mikroscopischen beobachtung", *Z. Tech. Phys., 16,* 454 (1935).

3. A. Marechal and P. Croce, "Un filtre de frequences spatiales pour l'amelioration du contrast des images optiques", *C.R. Acad. Sci., 127,* 607 (1953).

4. J. Tsujiuchi, "Correction of optical images by compensation of aberrations and by spatial filtering", *Progress in Optics, Vol. 2,* (E. Wolf, Ed.) North Holland Publishing Co., 133-180 (1962).

5. E.L. O'Neill, "Spatial filtering in optics", *Trans. IRE, IT-2,* 56-65 (1956).

6. A.W. Lohmann, " Optical single-sideband transmission applied to the Gabor microscope", *Opt. Acta, 3,* 97-99 (1956).

7. L.J. Cutrona, E.N. Leith, C.J. Palermo, and L.J. Porcello, "Optical data processing and filtering systems", *Trans. IRE, IT-6* 386-400 (1960).

8. L.J. Cutrona, E.N. Leith, L.J. Porcello and W.E. Vivian, "On the application of coherent optical processing techniques to synthetic-aperture radar", *Proc. IEEE, 54,* 1026-1032 (1966).

9. A. Kozma, E.N. Leith, and N.G. Massey, "Tilted plane optical processor", *Applied Optics, 11,* 1766-1777 (1972).

10. A.B. Vander Lugt, "Signal detection by complex spatial filtering", *Trans. IEEE, IT-10,* 139-145 (1964).

11. J.B. Breckinridge, Jet Propuslion Laboratory, California Institute of Technology - private communication.

12. B.J. Thompson, "Hybrid processing systems - an assessment", *Proc. IEEE, 65,* 62-76 (1977).

13. David E. Glover, "An Optical Fourier / Electronic Neurocomputer Automated Inspection System", *Proceedings of the IEEE International Conference on Neural Networks,* pp. I-569 - I-576, (1988).

14. B. Braunecker, R. Hauk, and A.W. Lohmann, "Optical character recognition based on nonredundant correlation measurements", *Applied Optics, 18,* 2746-2753 (1979).

15. C.F. Hester and D. Casasent, "Inter-class discrimination using synthetic discriminant functions (SDFS)", *Proc. SPIE, Vol. 302,* 108-116 (1981).

16. A.H. Rosenthal, "Application of ultrasonic light modulation to signal recording, display, analysis, and communications, *Trans. IRE, UE-8,* 1-5 (1961).

17. L. Slobodin, "Optical correlation technique", *Proc. IEEE, 52,* 1782 (1963).

18. M. Arm, L. Lambert and I. Weisman, "Optical correlation technique for radar pulse compression", *Proc. IEEE, 52,* 842 (1964).

19. W.T. Rhodes, "Acousto-optic signal processing: convolution and correlation", *Proc. IEEE, 69,* 65-79 (1981).

20. A.P. Goutzoulis, I.J. Abramovitz, "Digital electronics meets its match", *IEEE Spectrum, 25,* No. 8, pp. 21-25, August 1988.

21. N.J. Berg and J.N. Lee, *Acousto-optic Signal processing: Theory and Implementation,* Marcel Dekker, New York, 1983.

22. R.M. Montgomery, "Acousto-optical signal processing system", U.S. Patent 3634749, January 1972.

23. R.A. Sprague and C.L. Koliopoulis, "Time integrating acousto-optic correlator", *Applied Optics, 15,* 89-92 (1975).

24. P. Kellman, "detector integration acousto-optic processing", *Proc. 1978 Int. Optical Computing Conf., (Digest of Papers),* IEEE Order No. 78CH-1395-2C, 91-95 (1978).

25. T.M. Turpin, "Time integrating optical processors", in *Real time signal processing,* (F. Tao, Editor), *Proc. SPIE, 154,* 196-203 (1978).

26. R.P. Bocker, "Matrix multiplication using incoherent optical techniques", *Applied Optics, 13,* 1670-1676 (1974).

27. K. Bromley, "An incoherent optical correlator", *Optica Acta, 21,* 35-41 (1974).

28. M.A. Monahan, K. Bromley, and R.P. Bocker, "Incoherent optical correlators", *Proc. IEEE, 65,* 121-129 (1977).

29. J.W. Goodman, A.R. Dias, L.M. Woody and J. Erickson, "Some new methods for processing electronic image data using incoherent light", in *Optica Hoy y Manyana, Proc. ICO-11,* 139-145, (J. Bescos, A. Hidalgo, L. Plaza and J. Santamaria, Eds.), Sociedad Espaniola de Optica, Madrid, Spain (1978).

30. J.W. Goodman, A.R. Dias and L.M. Woody, "Fully parallel, high-speed incoherent optical method for performing discrete Fourier transforms", *Optics Let., 2,* 1-3 (1978).

31. D. Psaltis, D. Casasent, and M. Carlotto, "Iterative color-multiplexed electro-optical processor", *Optics Let., 4* 348-350 (1979).

32. H.J. Caulfield, D. Dvore, J.W. Goodman, and W.T. Rhodes, "Eigenvector determination by iterative optical methods", *Applied Optics, 20,* 2263-2265 (1981).

33. J.W. Goodman and M. Song, "performance limitations of an analog method for solving simultaneous linear equations", *Applied Optics, 21,* 502-506 (1982).

34. R.A. Athale and W.C. Collins, "Optical matrix-matrix multiplier based on outer product decomposition", *Applied Optics, 21,* 2089-2090 (1982).

35. H.T. Kung and C.E. Leiserson, "Algorithms for VLSI processors", in *Introduction to VLSI Systems*, C.A. Mead and L.A. Conway, Addison-Wesley, Reading, Mass., 271 (1980).

36. H.J. Caulfield, W.T. Rhodes, M.J. Foster, and S. Horvitz, "Optical implementation of systolic array processing", *Optics Comm., 40,* 86-90 (1982).

37. D. Casasent, "Acoustooptic transducers in iterative optical matrix-vector processors", *Applied Optics, 21,* 1859-1865 (1982).

38. M. Tur, J.W. Goodman, B. Moslehi, J.E. Bowers, and H.J. Shaw, "Fiber-optic signal processor with applications to matrix-vector multiplication and lattice filtering", *Optics Let., 7,* 463-465 (1982).

39. P.N. Tamura, P.R. Haugen, B.K. Betz, "Time integrating digital correlator", *Proc. SPIE, Vol. 431,* 121-126 (1983).

40. A. Huang, Y. Tsunoda, J.W. Goodman, and S. Ishihara, "Optical computation using residue arithmetic", *Applied Optics, 18,* 149-162 (1979).

41. W.T. Rhodes and P.S. Guilfoyle, "Acoustooptic algebraic processing architectures", *Proc. IEEE, 72,* 820-830 (1984).

42. A.P. Goutzoulis, D.K. Davies and E.C. Malarkey, "Prototype position-coded residue look-up table using laser diodes", *Optics Comm., 61,* 302-308 (1987).

OPTICAL COMPUTING WITH OPTICAL SPATIAL LIGHT MODULATORS

Stuart A. Collins, Jr.

The Ohio State University, ElectroScience Laboratory,
Columbus, Ohio 43212 U.S.A.

1.

INTRODUCTION

The optical spatial light modulator has been one of the main elements that has raised optical computing from the realm of novelty to that of usefulness. Basically it is an optical image amplifier, possibly nonlinear, capable of having, for example, dim images made in incoherent light as input and bright images made in coherent light as output. This has made it indispensable as a real time incoherent to coherent light converter for the Fourier transform type of optical computing and as a multichannel optical amplifier for numerical optical computing. It is such types of optical computing that we wish to consider here.

The generic form of an optical spatial light modulator, (OSLM) is shown in Fig. 1a where we see a black box with a light beam, which we'll call the write beam, containing an optical image incident on one side. On another side is a second brighter beam, the read beam, reflected off it. The image on the write beam is impressed on the read beam; this is the basic operation. How this is accomplished depends on the construction of the OSLM and will be discussed in detail later. The read beam may also be in some cases transmitted through the device, as in Fig. 1b.

Some spatial light modulators control the read beam using polarization
control. The read beam starts with linearly polarized light and the OSLM
changes it to elliptically polarized light, circularly polarized light, or light linearly polarized perpendicular to the initial read beam, depending on the write
beam intensity. The polarization control can be used by itself, or, if used with
an analyzer, it gives input-output intensity characteristic curves that can
either saturate or oscillate, depending on the OSLM. The curve can either
start from zero, or maximum, depending on the analyzer orientation.

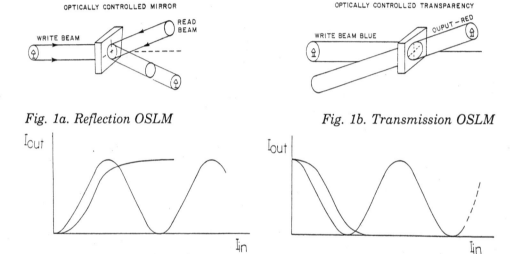

Fig. 1a. Reflection OSLM   Fig. 1b. Transmission OSLM

Fig. 1c. Characteristic input-output curves

Note that there can also be a negative gain characteristic where the read
beam gets dimmer when the write beam gets brighter. That can be especially
useful because there are various devices such as oscillators and stabilized am-
plifying systems that require negative gain amplifiers.

In our discussion we will talk about using the OSLM for various types of
optical computing. On one hand we will touch on the older Fourier-transform
optical computing based on the ability of a lens to perform a two-dimensional
Fourier transform operation. On the other hand we will present numerical
optical computing where we think in terms of number crunching and logic
operations similar to those used in electronic computers. This will again be divided into two types depending on the arithmetic used. We will also review the
various types of spatial light modulators currently commercially available and
will finish with some more sophisticated optical circuits.

2.
FOURIER TRANSFORM-BASED COMPUTING

We start by reviewing the first type of optical computing demonstrated, namely Fourier transform optical computing [1]. Its basic operation is shown in Figure 2 where we see an OSLM with the write beam formed from a scene. The read beam uses laser light which has the scene impressed onto it.

The OSLM is placed in the left focal plane of a lens. In terms of the transverse coordinates x,y the functional variation of the amplitude $E_2(x_2,y_2)$ of the optical wave in the right hand focal plane of the lens is the two-dimensional Fourier transform of the left hand focal plane field, $E_1(x_1,y_1)$ as shown in Figure 2. The right hand focal plane, affectionately referred to as the spectral or transform plane, has coordinate distances x_2,y_2 related to spatial frequencies κ_x and κ_y by $\kappa_x=2\pi x_2/\lambda f$, and $\kappa_y=2\pi y_2/\lambda f$.

The Fourier transform operation becomes more useful when one adds another lens to perform the inverse Fourier transform as shown in Figure 2. and puts spatial filters in the spectral plane. Typical applications include highlighting, where lines in a given direction or objects of a particular shape are highlighted, and matched filtering, where everything but objects of a particular shape is rejected and the objects replaced with a bright spot.

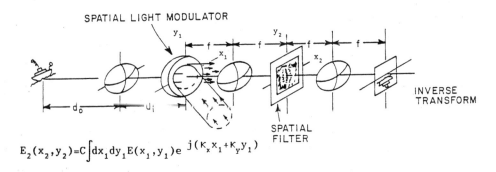

$$E_2(x_2,y_2)=C\int dx_1 dy_1 E(x_1,y_1)e^{j(\kappa_x x_1+\kappa_y y_1)}$$

Fig. 2. Complete real-time optical spatial filtering apparatus using OSLM

In addition to using an OSLM as an image transducer or a real-time incoherent-to-coherent light converter an OSLM can also be regarded as a a set of image amplifiers with as many independent channels as image elements. That leads us to the subject of numerical optical computing.

3.
NUMERICAL OPTICAL COMPUTING BASICS

We now consider some of the fundamentals of numerical optical computing. Two types of arithmetic will be used, binary and residue arithmetic. First we'll use binary arithmetic with its associated logic gates and flip-flops and give a simple optical circuit configuration using these basic elements. Then we will introduce residue arithmetic and later give an associated configuration.

Binary Computing

We now continue on to the magic world of optical binary computing with optical spatial light modulators. We will describe schemes using the OSLM to form both logic gates and flip-flops, then present one simple binary circuit.

Logic gates

One scheme for setting up the logic gates is shown in Figure 3a[2]. There we see an OSLM with input light beams impinging on two image elements and one light beam reflected successively off the corresponding output positions. Input light of one intensity is used. At that intensity the OSLM rotates the polarization plane by ninety degrees. With no input light the plane of polarization remains unchanged. In Fig. 3a the top and bottom polarizers are oriented so as to pass light which is polarized in the plane of the paper. The second polarizer is oriented to pass light perpendicular to it.

The only way to have a bright output is to have bright inputs at both X and Y to rotate the planes of polarization appropriately. If we assume bright true logic where bright represents a logical *one* and darkness represents a logical *zero* then we have the truth table shown, just that of a logical AND gate.

An alternative configuration for an AND gate is shown in Fig. 3b. There the read beam is taken to be input Y, and input X is applied to the write side of the OSLM. The polarizer passes light in the plane of the paper and the analyzer passes light perpendicular to it. Again, in order to have a bright output there must be inputs at both X and Y.

If the polarizer in the center is removed, giving the configuration shown in Fig. 3c, the truth table shown there results; that of a logical EXCLUSIVE OR gate.

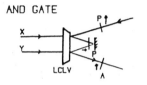

X	Y	OUT
D	D	D
D	B	D
B	D	D
B	B	B

*Fig. 3a. AND gate with
two write spots*

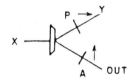

*Fig. 3b. AND gate with
single write spot*

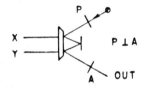

X	Y	OUT
D	D	D
D	B	B
B	D	B
B	B	D

Fig. 3c. Exclusive OR gate

Other logic gates are formed by the same configurations but with different polarization orientations and by using dark true logic[2] where a bright beam represents a *zero* and dark represents a *one*.

Binary adder

Now consider a simple optical adder constructed from optical AND and EX-CLUSIVE OR gates [3]. The basic operations are shown in Fig 4. It is well known from switching circuit theory that the sum and carry operations of a single digit of binary numbers X and Y can be formed as shown in Fig. 4a. The optical equivalent of the sum operation for that one digit is shown in Fig. 4b, being composed of cascaded EXCLUSIVE OR operations. The optical manifestation of the carry for one binary bit has three AND gates all feeding a three input EXCLUSIVE OR gate as shown in Fig. 4c.

The combination of the sum and carry operations to form a full adder is then given in the top part of Fig. 4d. The same optical circuit is repeated in levels parallel to the paper, there being as many levels as there are binary digits in the numbers. The carry from the i-th binary digit is side-stepped to go to the (i+1)-th digit. The output is shown at the center right.

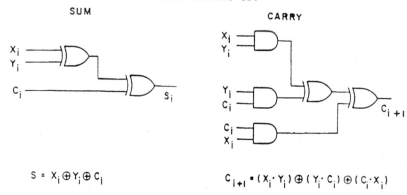

$$S = X_i \oplus Y_i \oplus C_i$$

$$C_{i+1} = (X_i \cdot Y_i) \oplus (Y_i \cdot C_i) \oplus (C_i \cdot X_i)$$

Fig. 4a. Electronic sum and and carry circuit

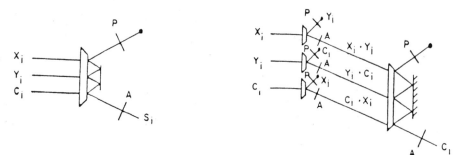

Fig. 4b. Optical sum configuration   *Fig. 4c. Optical carry configuration*

Flip-flops

Optical flip-flops can be formed by feeding the output of an OSLM back onto itself [4]. Two schemes will be presented, one involving two image elements per
flip-flop and giving complementary outputs, and the other using only one
OSLM image element and giving only one output.

The first scheme is shown in Fig. 5. There we see at the top center and bottom center of the page two OSLM's. The OSLM's are configured with appropriate analyzers so that a dark input gives a bright output past the analyzer
and vice versa. The output of the top OSLM is imaged onto the input of the bottom one and the output of the bottom one is imaged onto the input of the top
one.

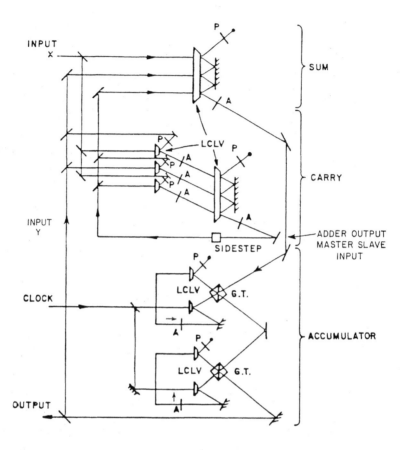

Fig. 4d. Optical temporal integrator circuit

This forms a nice self-consistent situation so that a dark input to the top OSLM generates a bright output which is imaged onto the input of the bottom OSLM, the output of which is then imaged onto the input of the top one, providing the self-consistent dark input. A second self-consistent configuration can be formed by interchanging dark and bright, giving the two states of the flip-flop.

In actuality the OSLM's shown in Fig. 5 represent two image elements of a single OSLM. There are possible as many as half the number of flip-flops as there are image elements. For an OSLM built roughly to television specifications, i.e. with six hundred lines and six hundred image elements per line, this could mean as many as one hundred and eighty thousand ($600^2/2$) independent flip-flops accessed in parallel.

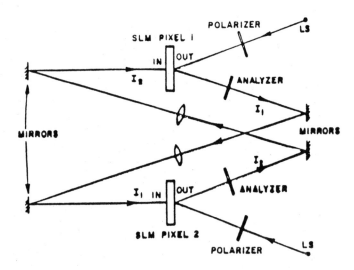

Fig. 5. Two OSLM spot flip-flop

The second method of generating flip-flops is shown schematically in Fig. 6 [5]. There we see again an OSLM at the top with the output imaged by means of two lenses onto the input. The OSLM and polarizers are adjusted so that a bright input at a given image element produces a bright output and vice versa. The two lenses are adjusted so as to produce a magnification of +1 so that each image element on the read side is imaged onto the corresponding image element on the write side. Assuming the same approximate television format, one OSLM has the capability of supporting three hundred and sixty thousand (600^2) independent flip-flops of this type, one for each OSLM independent image element.

Clocked flip-flop

Next we present a more elegant type of flip-flop, namely a clocked flip-flop; that is, one that has clock pulses applied and can be switched to a given state by means of an external beam when the clock pulse is on and holds the state when the clock pulse is off.

The clocked flip-flop is based on the second or single element flip-flop, redrawn somewhat in Fig. 7. The read beam is polarized in a plane parallel to the paper and the OSLM is configured so that with a bright write beam at a given image element the read beam has its polarization rotated ninety degrees. There is a polarizing prism which reflects light polarized perpendicular to the plane of the paper and transmits light in the plane of the paper. Thus when the flip-flop is in the bright state, read light is reflected by the polarizing prism and passed by the analyzer.

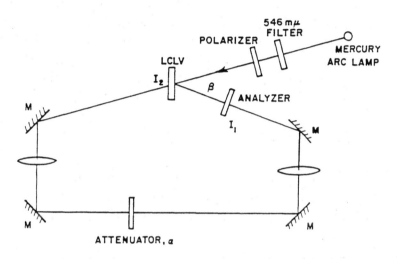

Fig. 6. Single OSLM spot flip-flop

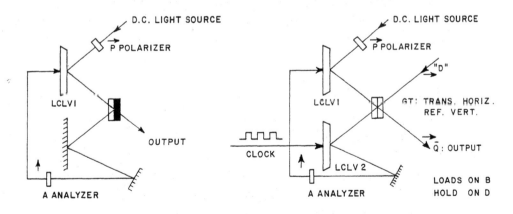

Fig. 7. Single OSLM spot flip-flop *Fig. 8. Clocked flip-flop*

The full clocked flip-flop is shown in complete detail in Fig. 8. There the mirror below the OSLM has been replaced by a second OSLM, or at least a second OSLM image element, which we'll call the clock element. Also, an input beam polarized horizontally so that it is transmitted through the polarizing prism from the upper right, has been added. With no write beam input to the clock element the read beam light is reflected off it with no polarization change and the flip-flop operates in whatever state it was initially set.

When the clock pulse is turned on a new state can then be entered into the flip-flop with the beam from the upper right passed by the polarizing prism. The horizontally polarized light transmitted by the prism has its polarization rotated by the OSLM clock element so that it is transmitted through the analyzer to the flip-flop write beam input. When the clock pulse is subsequently turned off, the read light from the flip-flop element is then passed through the system to the flip-flop input; thus maintaining the state.

Master-slave flip-flop

We can now combine two clocked flip-flops to form a master-slave flip-flop where a pulse is fed into the master portion of the master-slave and then passed onto the slave for safe keeping while the master is reading in a new pulse.

The master-slave flip-flop is shown in the bottom half of Fig. 4d. There we see two clocked flip-flops with the output of one on the top, the master, serving as the input to the other, the slave. There are two differences between the master and slave: the polarizer in the slave is oriented at ninety degrees to that in the master and the slave reads in with clock pulse off.. The net result is that the slave holds a state and the master reads in one when the clock pulse is on and vice versa.

Optical binary temporal integrator

Optical binary temporal integrator is a catch phrase for a device that takes in a sequence of numbers and keeps a running sum. The operation is illustrated schematically in Fig. 9. As shown we need the adder (on the left) and an accumulator formed from two clocked flip-flops, so a new state is not being stored at the same time as an old one is being read out.

At the beginning of operation the clock pulse is on and the accumulator is initialized to zero. The zero is added to the first term in the sequence and fed into the master flip-flop. The clock pulse is then switched off and the number is fed to the slave flip-flop. When the second number in the sequence arrives, the clock is turned on and it is added to the output of the slave and is stored in the master flip-flop, etc.

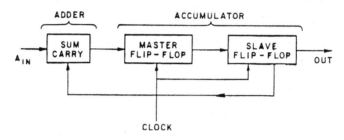

Fig. 9. Temporal integrator block diagram

The full optical circuit is shown in Fig. 4d. The connections shown correspond to only one digit. The optical circuitry for other digits is in planes parallel to the plane of the paper, above and below it. At the top of the figure we see the adder with its sum and carry, and at the bottom are the master and slave flip-flops. The input for the sequence of numbers is at the upper left, the clock input is at the left, a bit below center, and the running sum is at the lower left.

The time required to add one term in the sequence of numbers is measured in OSLM response times. One response time is needed for each binary bit in the addition operation while the clock pulse is on and one more response time is required to shift the number from the master to the slave flip-flop giving (n+1) response times for an n bit binary number.

The preceding has been an indication of the potential simplicity of forming optical circuits. The approach was basically to design the circuit using electronic techniques and then to transpose it to the optical manifestation.

Residue Arithmetic

Having given an example of an optical configuration using binary arithmetic we now go on to consider a different arithmetic approach, namely residue arithmetic [6]. Residue arithmetic is a parallel arithmetic which avoids the necessity of the ripple carry found in simple binary adders.

In residue arithmetic one deals with integers. One starts with a set of moduli, sometimes called bases, usually relatively prime numbers. Then the residue of a particular number with respect to one of the moduli is the remainder after dividing the number by the modulus. Any given number is represented by the set of residues with respect to the set of moduli. The representation is unique in the range equal to the product of the moduli if they are relatively prime.

This is illustrated in Table I where the chosen moduli 2,3, and 5 are given across the top. Numbers from 0 to 30 are shown at the left with a few omitted for brevity. The residues are shown in the columns under the respective moduli. For example, with these moduli the representations of 9 and 3 are (1,0,4) and (1,0,3), respectively.

Addition is performed by independently adding the respective residues and casting out the particular modulus if necessary. Thus the sum of the representations of 9 and 3 is (2,0,7) which becomes, after casting out the moduli 2 and 5, (0,0,2). Referring to Table I shows this to be the representation of 12 as desired. Subtraction is performed by subtracting residues or by adding a complement. Multiplication is performed by multiplying the individual residues and casting out the moduli. Thus the product of 9 and 3 is (1,0,2) which is the representation of 27. Division is more difficult but can be performed by techniques based on continued subtraction.

The choice of moduli is important. One wants to achieve an optimum utilization of the the light valve image elements consistent with computational ability. A typical choice of moduli might be nineteen, twenty, and twenty-one. These are relatively prime and give a dynamic range of eight thousand, roughly thirteen binary bits. That requires sixty image elements in the input plane and sixty elements in the output plane. One can do somewhat better with a sequence of smaller moduli such as 2,5,7,9, and 13 (dynamic range of 8190 using 36 image elements) but with somewhat increased complexity.

A residue arithmetic unit will be shown later, namely a matrix vector multiplier using a mapping approach to residue arithmetic operations.

4.

OPTICAL SPATIAL LIGHT MODULATORS

Now that we have considered some of the basics of numerical optical computing with an optical spatial light modulator we need to take a look at some typical commercially available spatial light modulators.

p=	2	3	5	
z	r_2	r_3	r_5	$z^{-1}\text{Mod}5$
0	0	0	0	-
1	1	1	1	1
2	0	2	2	3
3	1	0	3	2
4	0	1	4	4
5	1	2	0	-
6	0	0	1	1
7	1	1	2	3
8	0	2	3	2
9	1	0	4	4
10	0	1	0	-
11	1	2	1	
12	0	0	2	
27	1	0	2	
28	0	1	3	
29	1	2	4	
30	0	0	0	

```
ADD              9    1 0 4
               + 3    1 0 3
                12    0 0 2

SUBTRACT         9    1 0 4
                -3    1 0 8
                 6    0 0 1

MULTIPLY         9    1 0 4
                 3  X 1 0 3

PRODUCT =27: 1  0  2
```

Table 1

Liquid Crystal Light Valve

The Liquid Crystal Light Valve(LCLV)[7, 8] is made by Hughes and uses a cadmium sulfide photoconductor and a liquid crystal sheet as the light controlling agent. The cross section of an LCLV is shown in Fig. 10a where we again see a sandwich type of construction. The outer layers on both sides are indium tin oxide transparent electrodes. Next to the transparent electrode on the left is the CdS photoconductor followed by a CdTe light blocking layer. Next to the transparent photoconductor on the right is the liquid crystal layer followed by a multilayer dielectric mirror.

In operation the CdS-CdTe layer forms a back-biased diode which acts as a controllable capacitor. The liquid crystal cell also acts as a fixed capacitor in series with the back-biased diode capacitor. With no light injected the capacitance per area of the back-biased diode capacitor is small compared with that of the liquid crystal cell and most of the voltage is across the back-biased diode. Shining light on the CdS increases the available charges and the capacitance per area of the back-biased diode which decreases the fraction of the voltage across the back-biased diode and increases that across the light-controlling liquid crystal layer.

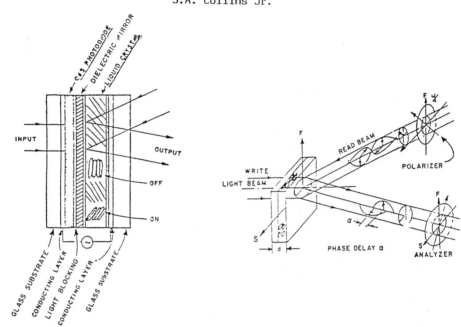

Fig. 10a. LCLV - cross section   Fig. 10b. LCLV - operation

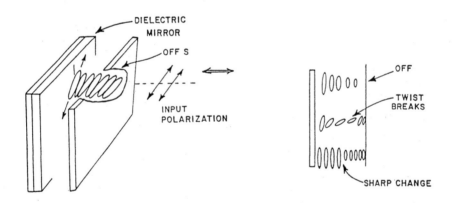

Fig. 10c. Twisted nematic   Fig. 10d. Twisted nematic
configuration - no voltage   configuration - operation

The liquid crystal layer controls the light by means of birefringence. The liquid crystal molecules being long and thin as they are, naturally are birefringent. The amount of birefringence depends on the orientation of the molecules which is in turn controlled by the voltage across the cell and by the cell walls. The liquid crystals are of the nematic variety; the natural state is to be parallel to each other. The cell walls are treated, for example, by rubbing them, to give molecules next to them a preferred direction.

In one simple configuration the liquid crystal molecules are, with no voltage, all aligned parallel to each other as shown in the center of the cell in Fig. 10a. The application of a voltage induces a dipole moment in the molecules which in turn causes them to tip to be more parallel to the associated electric field and perpendicular to the cell wall. In the process the refractive index for light polarized along the molecular axis $n_s(V)$ decreases while that perpendicular to the molecules nf remains constant.

In operation the read beam is polarized at forty-five degrees to the fast and slow birefringence axes as shown in Fig. 10b and there is an analyzer crossed with the polarizer. There is a delay, α, in electromagnetic phase between optical polarizations along the birefringent fast and slow axes given by the expression $\alpha = 2\pi(n_s(V)-n_f)2d/\lambda$ where d is the cell thickness and λ is the free space wavelength. The intensity out of the analyzer in Fig. 10b is given by $I = I_0 \sin^2\alpha$ so that the intensity oscillates, finally going to zero.

This periodic dependence of output intensity on input intensity gives rise to the oscillating characteristic such as that in Fig. 1c. At the point where the output intensity is maximum, $\alpha=\pi$, the polarization direction has been rotated ninety degrees. This configuration is especially useful for logical EXCLUSIVE OR gate if write beam intensity is controlled to give no more than a half-wave shift. it is also useful for for logical AND gates and other gates.

In another configuration, a twisted nematic configuration [9] shown in Figs. 10c and d, the preferred direction of the molecules at the cell walls are at an angle of forty-five degrees to each other and with no applied voltage the molecules exhibit a slow twist from one wall to the other.

The light of the entering read beam is initially set to be parallel to the nearest substrate and follows the twist of the molecules in to the mirror, is reflected, and follows the twist back, leaving the cell with the same orientation at which it entered.

On the other hand, with moderate voltages all the molecules in one half of the cell are partially tipped but mainly parallel to the preferred direction of one cell wall while those in the other half are also partially tipped but mainly parallel to the direction of the other wall. In the center there is a rapid change in orientation.

For that case the light enters with a polarization parallel to the molecules in that half of the cell; it continues through the first half of the cell with no polarization change until it comes upon the molecules in the other half of the cell which are at forty-five degrees to the polarization direction. The thickness of the cell is such that by the time the light has traversed the second half of the cell, been reflected off the dielectric mirror and traversed back through that half of the cell it has experienced a half wave phase shift and a ninety degree rotation of polarization. It then passes out through the near side with no further polarization change, ending up with polarization perpendicular to the entering direction.

Thus, in this configuration as the voltage increases, the polarization of the exiting read beam is parallel to the input direction for low voltages and ends up perpendicular to it for higher voltages. The intensity of the light passed by a crossed analyzer goes from zero to maximum. This explains the saturating behavior of the second curve in Fig. 1c. This configuration can give up to 100:1 extinction ration and can provide a sharp cutoff useful for making logical AND and OR gates.

The LCLV has a resolution of somewhat better than six hundred lines with six hundred points per line, input sensitivity is $\sim 100 \ \mu W/cm^2$, and the response time is approximately fifty milliseconds.

Pockels Readout Optical Memory

The Pockels Readout Optical Memory or PROM [10, 11] as it is called, is indeed truly an optical memory. A pattern is read in and remains there until erased.

The heart of the device is a thin crystal of bismuth-silicon-oxide, $Bi_{12}SiO_{20}$ affectionately referred to as BSO, which is simultaneously photoconductive and birefringent. The cross section is shown in Fig. 11 where we see a sandwich type of construction. On the outside are two transparent electrodes; inside each of the transparent electrodes is a layer of dielectric and in the center is the BSO crystal.

The device operates by applying a high voltage to the electrodes to create a strong electric field across the BSO, making it birefringent. A pattern is read in using blue light which causes the BSO to become photoconductive in areas where the blue light is bright. This in turn causes charge to leak off, thereby reducing the electric field across the BSO and reducing the birefringence at those areas. The pattern is then read out using red light polarized at forty-five degrees to the fast and slow birefringence axes as was done with the parallel configuration of the liquid crystal light valve.

In practice a somewhat more complex scheme is used in which one reverses the voltage thereby doubling the voltage across the BSO and achieving a larger effect for a source with a given voltage.

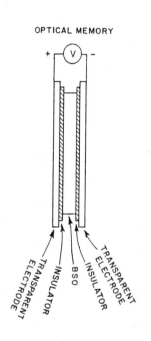

Fig. 11. PROM

The PROM works best with simple in/out operations because it requires light of different colors for the read and write operations. The operations can not be easily cascaded. The PROM is an integrating device, the effect being proportional to the number of joules per unit area in the write beam rather than the intensity of the write beam. Also high voltages of the order of a thousand volts are required. The resolution is around twenty-five cycles per millimeter, and the frame rate is approximately thirty frames per second. Four hundred ergs/square centimeter are required to reduce the output to ten percent of its original value.

Microchannel Spatial Light Modulator

The Microchannel Spatial Light Modulator (MSLM) [12,13] uses electrons from a standard photocathode multiplied in number by a microchannel plate and deposited on the surface of a slab of birefringent material to control the voltage across material and therefore its birefringence.

The cross section of the MSLM is shown in Fig. 12. The parts shown are mounted in a vacuum. The write beam enters through the window on the left hand face and proceeds to the photocathode mounted on the microchannel plate. With the high electron gain in the tilted channels of the microchannel plate there is a usable current existing its right side. The the charges in this current are accelerated by the applied potentials towards the birefringent plate.

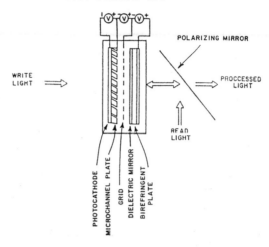

Fig. 12. MSLM - cross section

The resulting charges on the birefringent plate control point by point the voltage across the material and the resulting birefringence. The read beam is spatially modulated in the same way as in the parallel configuration of the liquid crystal light valve and the PROM. It passes through the birefringent slab and is reflected off a multidielectric mirror on the left side of the slab and passes back through the slab.

A typical readout configuration is shown: a polarizing beamsplitter reflects read light of one polarization and transmits light of the opposite polarization generated by the MSLM. With no birefringence there is no light transmitted.

There are several modes of operation depending on the relative voltages of the grid and the birefringent plate. In the first mode the grid pulls the electrons towards itself and past with such a velocity that they land on the slab and stick to it increasing the potential across the slab.

In an alternative mode the birefringent slab is first uniformly flooded with charge and then the grid voltage is increased to the point that the electrons arriving at the slab have sufficient energy to knock electrons off, thus decreasing the resident charge and its associated birefringence and producing the inverse image.

Alternatively, by adjusting the grid voltage properly thresholding and hard clipping are also possible. The thresholding capability leads immediately to the possibility of logical AND and OR gates depending on the threshold level.

As indicated there is a variety of operations depending on the associated voltages. However the same operation must be performed over the whole image plane. The device can be operated either in a real-time mode using a low resistivity birefringent slab so that the charges leak off or a high resistivity slab where the charges are stored. The voltages required are relatively high, a thousand volts or over. The resolution is typically around ten cycles per millimeter at fifty percent contrast level. The temporal response is around sixty hertz with fifty percent modulation depth. The input sensitivity is very good.

Other Spatial Light Modulators

There are two other spatial light modulators available which are not strictly optical spatial light modulators but which do warrant attention. They both are electronically addressed and require a TV camera. One, the Sight-Mod [14], which uses Ferrites to rotate the polarization plane has a 256x256 image elements array. The other is is the liquid crystal screen from a Radio Shack liquid crystal television [15]. It has 240 image elements vertically and 192 elements horizontally. For a beginning experimenter examining some new optical computing architecture with the intention of achieving results quickly and speeding up operation in the future the price is a good one.

5.

TYPICAL OPTICAL CIRCUITS

In this section we want to present some more complex examples of optical circuits. They include a design for an optical computer switching circuit and an optical residue matrix multiplier. We start with the optical computer switching system.

Optical Switching Network

For the first more complex example we will describe briefly a simple design for an optical switching network [16]. It is intended to interconnect in a simple fashion a large number (typically eighty) of sources such as computers or video devices requiring high bandwidth. It follows the protocols of a telephone switching system in that it looks for busy circuits before making a connection.

The network follows a star configuration as shown in Fig. 13. Between the central network and a given computer there are two connections to send and receive data and two connections to indicate the status of a call. For a simple design it is also assumed that there are as many additional lines as there are

other computers. It is assumed that the interconnections will be made using
optical fibers.

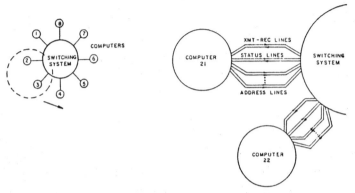

Fig. 13. Switching system configuration

An electronic switching diagram of the switching system is given in
Fig. 14. There we see the diagram partitioned into three subsystems; the status
subsystem, the clocking subsystem, and the holding subsystem. The status
subsystem checks to see if a given system is already connected, etc. before
making a connection. The holding subsystem makes and holds the data connection. The clocking subsystem puts in appropriate delays and activates the
holding subsystem when a connection is to be made or broken.

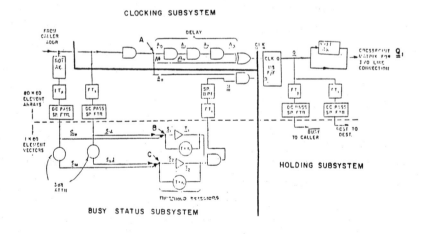

Fig. 14. Switching system electronic diagram

The actual device for making the data connections is shown in Fig. 15 where we see a typical matrix vector multiplier [17]. Data comes in from the various computers through the transmit (XMT) lines at the bottom and a portion of it is reflected into one of the receive (RCV) lines. The determination of which line is connected to which is made by the input to the LCLV which changes it from nonreflecting to reflecting.

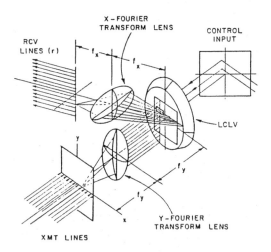

Fig. 15. Central switching element

The array of points for controlling the input to the connection matrix comes in through the array of address fibers as shown in Fig. 16. The lines from computer number one are lined vertically on the right side, those from computer number two are lined next to it, etc. If computer number eighty wants to communicate with computer number one then the fiber at the lower left of the array would be lit. The light from these fibers aids in turning on a master-slave flip-flop as shown in the bottom of Fig. 4d which stores the connection information.

The optical manifestation of the control system is shown in Fig 17. The master-slave flip-flop is in the holding system; the clock signal for the master-slave flip-flop comes from the address lines directly, while the setting information comes from the busy-status system which contains the logic for checking for an already connected line [18].

These serve to emphasize the possibility of more complex and useful computing systems using an OSLM.

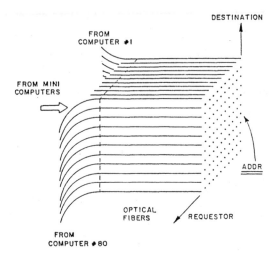

Fig. 16. Address fiber input array

Optical matrix-vector multiplier

The optical matrix-vector multiplier uses a Hughes liquid crystal light valve, residue arithmetic and a holographic lookup table [18]. It was designed to operate in minimum time, using one light valve response time, assuming that there was plenty of space available on the surface of the LCLV. In addition it uses binary position coding where the numbers are indicated by position along a line, and the numerical value is indicated by a bright spot at one location. It is also required that the output have the same coding as the input so that such devices can be cascaded.

The matrix-vector multiplier uses a particular representation for the residue arithmetic operations, that of mappings. These mappings determine the configuration of the unit and will be presented first.

We consider a simple arithmetic operation for a single modulus as shown in Fig. 18a. In the representation a given number for a particular modulus is represented by its position in the vertical array at the right. The numerical operation is represented by the square mapping and the resulting number is represented by its position in the horizontal array at the top. The mapping has one entry in each row and column. It corresponds to a shuffle where each element of the vertical input array is set equal to an element of horizontal output array.

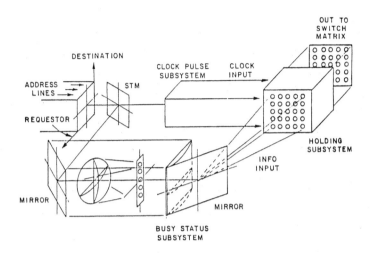

Fig. 17. Switching system optical configuration

One finds the connection by moving along the row until a one is found and then shifting up to the element of the associated output array element.

A different mapping is used to represent each arithmetic operation and to represent each numerical value of each operation. For example the mappings of *plus two* and *times four* both modulo 5 are shown in Figs. 18a and b, respectively. The *times four* modulo five mapping is easily explained: zero on the right times zero is still zero, one times four, modulo five is four, two times four is eight, and with a five cast out becomes three.

PLUS 2

$$
\begin{bmatrix}
0 & 0 & 1 & 0 & 0 \\
0 & 1 & 0 & 0 & 0 \\
1 & 0 & 0 & 0 & 0 \\
0 & 0 & 0 & 0 & 1 \\
0 & 0 & 0 & 1 & 0
\end{bmatrix}
\begin{bmatrix}
0 \\
1 \\
2 \\
3 \\
4
\end{bmatrix}
$$

Fig. 18a. Mapping for plus 2 mod 5

TIMES 4

$$
\begin{bmatrix}
0 & 0 & 0 & 0 & 1 \\
1 & 0 & 0 & 0 & 0 \\
0 & 1 & 0 & 0 & 0 \\
0 & 0 & 1 & 0 & 0 \\
0 & 0 & 0 & 1 & 0
\end{bmatrix}
\begin{bmatrix}
0 \\
1 \\
2 \\
3 \\
4
\end{bmatrix}
$$

Fig. 18b. Mapping for times 4 mod 5

In a residue representation with three moduli three such mappings of unequal size would be used: the input number being represented by three vertical arrays one above the other, and the output by three horizontal arrays next to each other.

In the matrix vector multiplier two physical units will be used: the first is an implementation of the operation, used in every row of the mappings, to shift left a given number of spaces and move to the output, the second is a holographic map generator. The construction and operation of these two units will now be described.

A simple conceptual implementation of the shift-left-and-out operation is shown in Fig. 19a for a single line of a mapping. There we see an LCLV on the left with write beams at two image elements. The LCLV is configured so that when there is no optical write beam input at a given image element, the read beam light will reflect off that element with no polarization change. With an optical input at that element the read beam has its polarization rotated ninety degrees.

In Figs. 19 we also see a polarizing mirror; it transmits light of one polarization and reflects light of the perpendicular polarization. This is a hypothetical device manufactured by the same company that produces frictionless massless pulleys for beginning physics experiments. However as will be shown, the same operations can be implemented using polarizing prisms and mirrors.

The configuration emulates the mapping: sidestep so many elements and move out. Light entering one of the input positions has its polarization rotated so that it is subsequently reflected by the polarizing mirror. After so many bounces it again, has its polarization rotated and is transmitted by the mirror. It then passes through a lens which focuses it down to the output line.

As mentioned, a number represented by three moduli would have three vertical arrays, one for each modulus. There would be three square areas on the LCLV and three horizontal output arrays. The particular arithmetic operation is determined by the LCLV's write beam.

A typical experimental manifestation implementing the polarizing mirror is shown in Fig. 20. The LCLV and its write beam input from a CRT are shown on the left and on the right is a plane mirror. On the top and bottom are two polarizing beam splitting cubes and next to them are two flat redirecting mirrors. Light polarized such that it is reflected from the cubes travels counterclockwise in a ring: from the LCLV to the bottom prism, to the right hand mirror, to the top prism, and back to the LCLV. Further, the redirecting mirror next to the top cube is oriented so that when the light arrives back at the LCLV it is at the spot adjacent to the one from which it started.

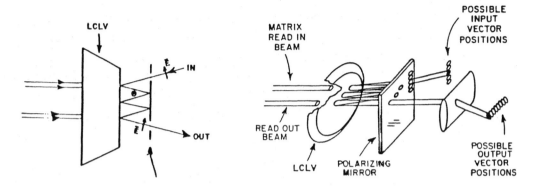

Fig. 19a. Conceptual "shift and out"
implementation

Fig. 19b. Optical "Shift and out"
implementation

To avoid possible diffraction broadening and to use the light valve to its fullest resolution each cube has lenses next to its faces so that the cube acts like a single thick lens of focal length, f. The LCLV and right hand mirror are separated from the polarizing beam splitters by optical distance of 2f so that the right hand mirror contains the image of the LCLV.

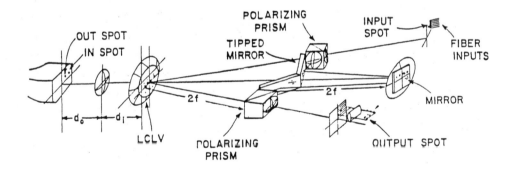

Fig. 20. Useable "shift and out" implementation

In operation, light comes into one of the five spots at the upper right, possibly through optical fibers. Light of the proper polarization passes through the top polarizing prism and is imaged on the light valve where its polarization is rotated to confine it to the loop. After n traversals around the loop, the light valve spot has moved over n image elements and strikes an element where its polarization is again rotated and it passes out through the bottom polarizing prism.

The second unit for implementing the residue mappings is a holographic lookup table. This is shown in Fig. 21. There we see a hologram and to its left, a lens. The hologram has a thick emulsion and several superimposed exposures, each reconstructed with a plane wave in a different direction. In the focal plane of the lens is an array of points; light in each point provides a plane wave in the proper direction to reconstruct a particular mapping pattern.

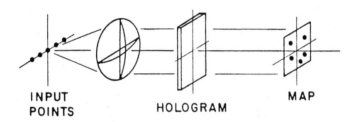

INPUT   M A P
POINTS   HOLOGRAM

Fig. 21. Holographic table lookup

The input points correspond to numerical values of the input. Each point reconstructs the mapping appropriate to its value, the arithmetic operation, and the residue modulus. For example, Fig. 21 corresponds to residue modulus five so the input points correspond to the numbers zero through four. If the arithmetic operation is addition, then the first point reconstructs the map for addition of zero, the second point reconstructs the map for addition of one, the third point reconstructs the map for addition of two, etc. These will be used to generate some of the write beam inputs to the LCLV for the shift-left-and-out operation.

The complete matrix vector multiplier performing the operation $\bar{c} = \bar{\bar{A}} \bar{b}$ or $c_j = \sum\limits_{i=1}^{N} a_{ji} b_i$ for one residue modulus and for one row of the mappings for that modulus is pictured schematically in Fig. 22. Just left of center is a tall thin looking LCLV with write beams provided by the image of a CRT. To its right are several polarizing mirrors performing the shift-left-and-out operation. To the left of center is another LCLV and set of polarizing mirrors. The input for this is provided by maps from holographic lookup tables.

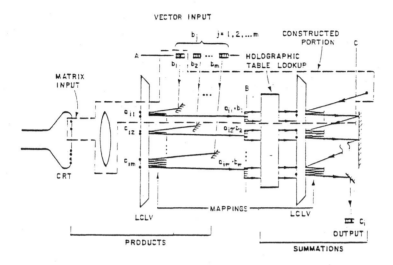

Fig. 22. Matrix vector multiplier optical schematic

In the Fig. 22 the left hand light valve and mappings provide multiplication by the numerical values of the matrix elements, a_{ji}. Write beam inputs for that light valve are generated by the CRT. The input elements b_i of the input vector are shown at the upper center. Each comes into its own polarizing mirror shift-and-out operation, multiplying it by the associated matrix element.

The output of the multiplication operations then goes to holographic lookup tables, each of which generates a mapping for the LCLV's write beam for the addition operations. The addition starts with an input of zero at the upper right, adding it successively to the values generated by the rest of the products. The final output is shown at the lower right.

To demonstrate feasibility a portion of the residue matrix-vector multiplier was constructed. The portion is that outlined by the dotted lines in Fig. 22 comprising one input vector element for one residue modulus, multiplication by a matrix element a_{ij} predetermined by a microprocessor-controlled CRT, a holographic lookup table and the first addition operation. The residue modulus chosen was five. The ring configuration was used for the shift-left-and-out operation. It was implemented twice on the same LCLV.

The implementation itself is shown in Fig. 23. The figure was prepared by making a line drawing from a photograph of the actual apparatus. The LCLV is shown just above center with the CRT for the matrix elements at the very top. The shift-left-and-out loops are in the center. The addition loop is below the multiplication loop. The polarizing prisms are of the Glan-Thompson type and labeled GT1 and GT2. In this implementation the loops were set up so that the light travels counterclockwise. The holographic lookup table is shown at the right.

In operation the input plane the vector element value for the multiplication loop is at the lower left. The light in the desired spot eneters GT1 from the lower left and, after the desired number of traversals, exits via GT2. From there the light goes to the holographic memory, the output of which forms thed write beam for the LCLV input corresponding to the addition loop. The LCLV read beam for the addition loop enters from the lower center and the value of the final output vector element is indicated by the position in the vector output plane.

As an example of the potential capability of this approach, using residue moduli 2,5,7,9,and 13 giving a dynamic range of a little over 2000, (thirteen binary bits) and a 2x2 matrix vector multiplication, one would have 36 elements in input and output vectors, $2^2+5^2+7^2+9^2+13^2=328$ image elements in multiplication tables for one matrix element, or 1312 image elements for all the multiplication operations. There would be an additional 1312 image elements required for the addition operations giving a total of 2624 required light valve image elements. For a light valve with 600x600 image elements in an area of 2.54^2 cm^2 this gives a potential of one hundred and thirty-seven 2x2 matrix-vector multiplies. With presently available LCLV this could be done in one LCLV response time of roughly fifty milliseconds. In the future, ten microseconds should be quite feasible with different OSLM's.

5.
ADDITIONAL WORK

There are several points worth mentioning in passing. In addition to the simple systems presented here there is also work demonstrated [19] and in progress [20] on general architectures and design aspects of OSLM-based optical computers. Binary computing using logic gates based on a sharp cutoff device are fit in well with binary electronic thinking. In addition there are other higher capability spatial light modulators under development [21, 22] and much work being done on architectures [23, 24, 25, 26, 27].

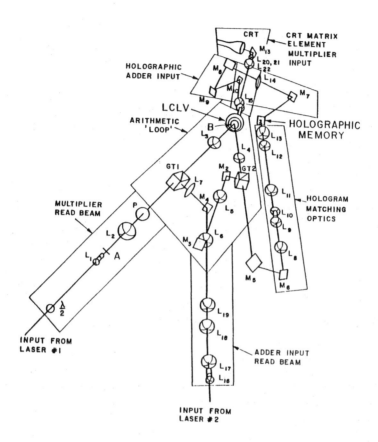

Fig. 23. Matrix vector multiplier optical implementation

6.

SUMMARY AND CONCLUSIONS

In this paper we have considered optical computing with optical spatial light
modulators. We have shown how optical logic gates and flip-flops can be
formed and how in principle they can be combined into optical circuits. We
have also considered residue arithmetic, useful because of its parallel nature
and have shown two typical optical circuits, an optical switching circuit and an
optical matrix-vector might be configured. We conclude that numerical optical
computing is indeed possible with OSLM's and is limited at present only by the
ingenuity and imagination of the designer.

ACKNOWLEDGEMENT

This work is sponsored in part by the National Aeronautical and Space Administration

REFERENCES

1 Goodman, J.W., "Introduction to Fourier Optics," Mc Graw Hill Book Company, New York (1968).

2 Fatehi, M.T., K.C. Wasmund, and S.A. Collins Jr., "Optical Logic Gates using a Liquid Crystal Light Valve," Applied Optics 20, 1424 (1981).

3 Fatehi, M.T., K.C. Wasmund, and S.A. Collins Jr., "Optical Flip-flops and Sequential Logic Circuits using a Liquid Crystal Light Valve," Applied Optics 23, No. 13, 2163 (1984).

4 Sengupta, U.K., U.H. Gerlach, and S.A. Collins Jr., "Bistable Optical Spatial Device using Direct Optical Feedback," Optics Letters, 3, 199 (1978).

5 Gerlach, U.H., U.K. Sengupta, and S.A. Collins Jr., "Single-SLM Bistable Optical Matrix Device using Optical Feedback, " Optical Engineering 19, 452 (1980).

6 Szabo, N.S. and R.I. Tanaka, "Residue Arithmetic and its Applications to Compute Technology," McGraw Hill, New York (1967).

7 Hughes Aircraft Co., Industrial Products Division, 6155 El Camino Road, Carsbad, CA 92008-4888.

8 Bleha, W.P., L.T. Lipton, E. Wiener-Avnear, J. Grienberg, P.G. Reif, D. Casasent, H.B. Brown, and B.V. Markevitch, "Applications of the Liquid Crystal Light Valve to Real-Time Optical Data Processing," Optical Engineering 17, No.4 (1978).

9 Fraas, L.M., J. Grinberg, W.P. Bleha, and A.D. Jacobson, "Novel Charge-Storage-Diode Structure for Use with Light-Activated Displays," Journal of Applied Physics 42, 2 (1976).

10 Itek Corp., Lexington, MA 02173.

11 Horowitz, Bruce A. and Francis J. Corbett, "The Prom-Theory and Applications for the Pockels Readout Optical Modulator," Optical Engineering 17, No. 4, 353 (1978).

12 Hamamatsu Co., 360 Foothill R., Box 6910 Bridgewatter, WI 08807.

13 Warde C., A.M. Weiss, A.D. Fisher, and J.I. Thakara, "Optical Information Processing Characteristics of the Microchannel Spatial Light Modulator, " Applied Optics 20, No. 12, 2066 (1981).

14 Semetex Corp., 3450 Fujita St., Torrance, CA 90505.

15 Liu, Hua-Kuang, Jeffery A. Davis, and Roger A. Lilly, "Optical-Data-Processing Properties of a Liquid-Crystal Television Spatial Light Modulator," Optics Letters 10, No. 12, 635 (1985).

16 Clymer, B. and S.A. Collins Jr., "Optical Computer Switching Network," Optical Engineering 24, No. 1, 74 (1984).

17 Goodman, J.W., A.R. Dias and L.M. Woody, Opt. Letters 2, 1 (1978).

18 Habiby, S.F. and S.A. Collins, "Implementation of a Fast Digital Optical Matrix-vector Multiplier using a Holographic Look-up Table and Residue Arithmetic," Applied Optics 26, 4639 (1987).

19 Jenkins, B.K., A.A. Sawchuk, T.C. Strand, R. Forchheimer, and B.H. Soffer, "Sequential Optical Logic Implementation," Applied Optics 23, No. 19, 3455 (1984),

20 Walker, "Application of Bistable Optical Logic Gate Arrays and All Optical Digital Parallel Processing," Applied Optics 25, 1578 (1986).

21 Walker, A.C., S.D. Smith, R.J. Campbell, J.G.H. Mathews, "Electron-beam-tunable Interference Filter Spatial Light Modulator," Optical Letter 13, 345 (1988).

22 Pape, Dennis R., "An Optically Adressed Membrane Spatial Light Modulator," proceedings on the Conference of Spatial Light Modulators and Application, Los Angeles, CA, January 26-27, 1984, Proceedings of the SPIE Vol. 465.

23 Applied Optics, 25, No. 10, 15 May, 1986, Digital Optics.

24 Applied Optics, 25, No. 14, 15 July, 1986, Optical Optics.

25 Topical Meeting on Optical Computing, Technical Digest Series 1987, Vol. 11, Optical Society of America, Washigton D.C. (1987).

26 Applied Optics 27, No. 9 (1988).

27 Optical Engineering, 25, No. 1 (1986).

OPTICAL SWITCHING DEVICES:
SOME BASIC CONCEPTS

D. A. B. Miller

AT&T Bell Laboratories, Holmdel, NJ 07733, USA

1.

INTRODUCTION

Optical switching devices, by which I mean devices with which light switches light, can take many forms. The many devices proposed can operate on many different principles and can perform many different functions. In this introductory summary, I will attempt to give a qualitative overview of this field. This will not be an exhaustive review, nor will it be a critical comparison of devices. Instead, I will try to explain briefly some of the concepts in the field, both in the physical principles and types of devices and in the requirements that systems impose on devices. I will also emphasize the importance of considering devices in the context of systems. It is important to realize that there is no "best" device; different systems require different device attributes.

Hopefully, in this article I will introduce much of the vocabulary and concepts in the field of optical switching devices. Space will not permit thorough discussions of any. I will start by discussing nonlinear optical processes in general, then I will review some physical attributes of devices, and finally I will summarize some systems requirements on devices.

2.

NONLINEAR OPTICAL PHENOMENA

For this discussion, I will take the broadest possible view on nonlinear optics; nonlinear optical effects are any phenomena that allow one light beam to affect another. Clearly such phenomena must be at the core of any optical switching device. I will treat such concepts as "real" and "virtual" transitions, local and nonlocal nonlinearities, nonlinearities that can be expanded as a power series in the optical field and those that cannot, all of which are relevant to optical switching devices.

What we could call "classical" nonlinear optics covers phenomena such as second harmonic generation and related wave mixing phenomena, and is based on the fact that, in all matter, the displacement of charge (i.e. the polarization) is not in general proportional to the applied field, especially when that field becomes comparable to the internal binding fields of the matter. Thus we can usefully think of this kind of effect in terms of nonlinear springs with charged masses attached; all harmonic generation and many other phenomena can be modelled this way. Important attributes of this kind of effect are as follows. (i) They are usually local, that is, fields at a given point only produce polarizations at that point; this is certainly not true on atomic distances, but is usually effectively true on length scales on the order of the light wavelength. (ii) They are usually fast, because the "resonant frequencies" of the springs are usually in the optical range (since they correspond to optical transitions in the material). (iii) They are weak, because fields of the order of the internal binding fields of matter are required for large effects. (iv) They rely on "virtual" transitions; once the springs are oscillating, the nonlinear process (such as second harmonic generation) can proceed without any dissipation in ideal lossless springs. Energy is however being exchanged, for example from the fundamental to the second harmonic, by being absorbed by the springs and coherently re-emitted. Quantum mechanically this proceeds through so-called "virtual" transitions. It is important to understand that such transitions do result in finite populations that are perfectly real in the general sense of the word, which correspond to the energy stored in the oscillating spring/mass system in our classical mechanical analog. The quantum-mechanical phase of these populations is however still coherent with the light field. Note that, although there is no steady state dissipation in such an ideal "virtual" system with monochromatic fields, there must be absorption of energy from the light field to start the springs oscillating in the first place. It is also in practice very difficult to recover all of that energy back into the light field when we are finished, and some fraction of it may be lost. As an illustrative analogy of this difficulty, if you drive your car, with no shock absorbers, over a bump, then some of your forward kinetic energy is converted into the vertical oscillations of your car. To convert this vertical energy back into forward kinetic energy requires a complementary bump of exactly the right size that is also positioned such that it catches your oscillation at exactly the right phase. (v) Such lossless systems are also intrinsically reversible in the thermodynamic sense (e.g. we can in principle run second harmonic generation backwards to generate the fundamental). This can be a problem when we desire irreversibility in our logic system.

As our optical frequency approaches a transition frequency, such virtual processes are resonantly enhanced, the time that the energy of an individual photon spends in the excited state of the matter becomes longer and longer (it is essentially given by the uncertainty time related to the detuning energy), and the probability that the quantum mechanical phase of the excited state will be interrupted by some random collision before re-emission becomes larger and larger. Eventually, this destroys the virtual transition physics, and we obtain "real" transitions, which can give absorption. It is the randomness of the collisions that makes the transition "real".

A typical real transition effect would be absorption saturation. This has received a lot of attention for optical switching, particularly in semiconductors, although it is more often the nonlinear refraction associated with this nonlinear absorption that is used for the switching. Here typically the absorption saturates because of the Pauli exclusion principle - the excited states become filled. The associated change in refractive index can be calculated from the Kramers-Krönig relations, which inescapably link absorption and refraction. The amount of intensity required for saturation can be low if the lifetime of the excited state is large, and there is generally a reciprocal relation between power and speed in such effects. We can sometimes approximately describe such effects in terms of nonlinear susceptibilities (e.g. the degenerate four wave $\chi^{(3)}(\omega,-\omega,\omega,-\omega)$), in which case we can obtain large numbers, but in doing so we must be aware of two potential problems: (i) because of diffusion of excitation (e.g. carrier diffusion), these effects can be non-local; (ii) the changes in properties are not actually induced directly by the field itself, but by the "degree of excitation" of the material (e.g. the carrier density), which in turn is dependent on the absorbed power. This latter effect has a non-trivial consequence, because it can lead to intrinsic bistability due to increasing absorption; if, for example, increasing carrier density gives increasing absorption (e.g. through band-gap renormalization) that in turn gives increasing carrier density and so on, we can have overall positive feedback and bistable switching. Such a bistable response of the material itself can never be described in terms of a power series in the internal optical field or intensity, because power series are always single-valued and bistability is triple-valued (there are two stable and one unstable values of the output for some range of values of the input). Another example of a real transition nonlinearity is a thermal nonlinearity; it also is non-local because of thermal conduction and in some cases can show bistability from increasing absorption. The "degree of excitation" in the thermal case is temperature.

Note that there is an important distinction between these intrinsic material bistabilities and bistabilities involving external feedback (e.g. Fabry-Perot bistability). In the latter case the whole system is bistable even although the response of the material (e.g. the polarization) is single-valued in the field inside the material, and hence can be described by a power series expansion.

Another example of non-local, non-power-series effects is provided by hybrid devices. Here we could electrically detect light in one place, communicate the resulting voltage to another, and use it through some electro-optic effect to change the optical transmission of another beam. It is also quite straightforward to make bistable systems, for example by incorporating electrical bistability between the detector and the electro-optic output device, or by some hybrid opto-electronic bistability as in simple self-electro-optic-effect devices (SEEDs) or bistable laser diodes, and hence the output of such systems cannot be described by a power series expansion in the field inside the photodetector.

In summary, it can be seen that there is a large variety of types of nonlinear effects that have been considered for use in optical switching devices. These effects have gone well beyond the "classical" nonlinear optics because of the necessity of providing devices operating with sufficiently low intensities, albeit at some compromise in speed of switching, and also partly because of the desire for sophisticated functionality and for behavior such as thermodynamic irreversibility in order to make system design easier.

3.

TYPES OF DEVICES

We can conveniently group many types of devices by means of opposite pairs. I will discuss several such pairs here: two-terminal v. three-terminal; bistable v. non-bistable; active v. passive; intrinsic v. hybrid; refractive v. absorptive. Between them, these categories include nearly all optical switching devices, and this discussion will also serve to introduce and define much of the associated terminology.

The terms "two-terminal" and "three-terminal" are simple to understand in their original electronic context: a two-terminal device, such as a tunnel diode, has only two connections, whereas a three-terminal device, such as a transistor, has three connections. Both of these types of devices can show gain and can switch. The real importance of the distinction lies in two qualitative attributes: (i) two-terminal devices themselves can make no distinction between input and output since both are connected to the same terminal (the other terminal is used for the ground or reference voltage for the signal in the electrical case), hence they have poor input/output isolation in that any fluctuations fed back into the output are liable to be amplified by the gain of the device; (ii) two-terminal devices usually suffer from critical biasing requirements because fluctuations in the biasing (e.g. power supply variations and load variations) also tend to be amplified by the device. Thus, for example, to exhibit signal gain a simple bistable device needs to be biased close to a switching threshold so that a small additional incident power can switch the device, making a large change in output power; this is obviously a critical biasing, and such a device can also be switched by small reflections into the output, showing poor input/output isolation. Three-terminal devices usually have neither of these attributes. In the optical context there is no direct analog of the "terminals" of electronic devices, but we can use a generalized definition in which "three-terminal" devices show input/output isolation and absence of critical biasing whereas "two-terminal" devices do not. In discussing input/output isolation, it is useful to consider the generalized "overlap" of input and output, with those devices with nearly "orthogonal" input and output (i.e. low overlap) having good input/output isolation. This kind of generalization is useful, for example, in discussing "time-sequential" gain as used in the recent symmetric-SEED devices; in this case, the input and output use the same physical connection, but when the device is sensitive to input it has little output, and when it subsequently has a large output, it has little input, so that the input and output have little overlap when the time integration is considered. Incidentally, it takes a three-dimensional graph to display the characteristics of a three-terminal device, whereas those of a two-terminal device can usually be displayed in two dimensions. The vast majority of three-terminal devices also utilize thermodynamic irreversibility to improve input/output isolation. Although three-terminal devices as defined here are obviously desirable for systems (as will be discussed below), it has proven difficult to make optical ones, and the majority of proposed devices are essentially "two-terminal" according to the present definition.

The most commonly investigated class of optical switching device is certainly bistable devices, that is devices that have two stable output states for a range of input powers. I discussed bistability from increasing absorption briefly in the previous section, which is a bistability that is intrinsic to the material. Most of the proposed bistability mechanisms, however, use a combination of a non-bistable optical nonlinearity with some external feedback. Of these, the most investigated system is Fabry-Perot nonlinear refractive bistability. In this system, the nonlinear Fabry-Perot resonator consists of a pair of plane, partially-reflecting mirrors surrounding a material whose index of refraction depends on the optical intensity within it. It is initially tuned slightly off resonance. With increasing input intensity, the intensity inside the resonator increases, thereby changing the refractive index of the cavity, and pulling it towards the resonant condition (in which an integral number of half wavelengths of light would fit inside the cavity). Because the cavity is now more nearly resonant, the intensity is magnified inside the cavity, giving yet further change in refractive index that pulls the system yet closer to resonance, thus further increasing the intensity inside the cavity and so on. This positive feedback mechanism, which results from the combination of the optical nonlinearity of the material with external feedback from the cavity, can become so strong as to switch the device into a highly resonant state. Once in this state, it can be maintained in a highly resonant condition with less incident light because the resonant cavity is magnifying the incident light intensity. Such a bistable system shows both hysteresis, in which the input/output characteristic is different for increasing and decreasing power, and also abrupt switching from one state to another at the switching powers. This bistability can be seen with many different materials and nonlinearities, and also with many different forms of cavity, thus giving a way of exploiting a large variety of materials and structures for optical switching. The resonance also reduces the incident intensity required for switching, therefore enhancing the usefulness of small optical nonlinearities. It has the disadvantages that it is basically still a two-terminal device as defined here and the cavity may have to be fabricated to very high precision, although this fabrication may be possible if it exploits highly-controllable layered growth technology. Many other bistable systems are possible by combining the general principle of nonlinearity combined with external feedback, although those not using resonators generally require higher intensities. Increasing absorption bistability does not normally take advantage of cavities, but can be interesting if used with very large nonlinearities, as in the case of SEEDs. In the case of SEEDs, three-terminal bistability is possible by making a device (the symmetric SEED) that is bistable in the ratio of two beam powers.

There is no fundamental system need for bistable devices, as will be discussed in the next section. Devices showing differential gain, in which small changes in one beam produce larger changes in another, could also be used as the basis for a switching system, and there are many ways of achieving such behavior. Many bistable devices can be adjusted to a regime where they are not quite bistable, but still show a strong "kink" in their input/output characteristics with a slope greater than one (i.e., differential gain), although again these are usually still two-terminal devices. Nonlinear Fabry-Perot resonators can be run with two different wavelengths, with light at one wavelength that is absorbed controlling the optical length of the cavity; hence much larger powers at a second wavelength where the material is transparent can be switched by moving a transmission resonance through this second wavelength. To make this system cascadable (see the next section) requires a complementary device that can be run with the wavelengths interchanged. Laser diode amplifiers can obviously give differential gain, although they can have input/output isolation problems since they amplify just as well in one direction as in the other. Opto-electronic systems including transistors can also obviously be set up to show differential gain without bistability.

A common conceptual difficulty is understanding how "passive" devices, that is devices which are not themselves a source of optical energy, can provide any kind of optical gain. "Active" devices, such as laser diode amplifiers or a device such as a phototransistor driving a light emitting diode, can clearly provide real optical gain since they convert some other source of power (e.g. electricity) into light. The key of course is that optical "passive" devices, just like the transistor in electronics, do only dissipate optical power, but can still induce large changes in power in one beam with small changes in power in another; such passive devices do require some external optical power source to generate the beam that will be modulated, but it may be possible to run many such passive devices off a single light source. (Note incidentally that the use of "active" and "passive" here is different from that in electronics, where "active" devices are components such as transistors, and "passive" refers to components such as resistors and capacitors.)

Another common distinction in switching devices is between "intrinsic" or "all-optical" devices and "hybrid" devices. There is some variability of meaning of these terms. "Intrinsic", when used in the sense of "not hybrid", usually means that there is no electronic transport (i.e. no movement of electrical currents), whereas "hybrid" means essentially opto-electronic in that some combination of optics and electronics is used. It should however be noted that the "classical" nonlinear optics actually involves coherent electrical transport in that the charge "clouds" are displaced by the field, and that this is the source of the nonlinearity itself. Perhaps "hybrid" therefore should be used only for those devices involving dissipative electronic transport. Hybrid devices have a reputation for being inefficient; it is a common opinion, for example, that it is inefficient to convert from optics to electronics and back to optics, and hence that the performance of such a device is fundamentally limited. In fact, the efficiency of hybrid devices depends on the degree to which they are integrated; there is nothing fundamentally very inefficient about photodetectors, laser diodes or some modulators (e.g. quantum well devices). The inefficiency comes about in practice from the energy required for electrical communication over macroscopic distances from one device to another; it must however be stated that the integration of all such components is a technologically very demanding task. We could however look upon "classical" nonlinear optical effects as being in some senses the ultimate integrated opto-electronic device since they are totally efficient in their optical-to-electronic-to-optical conversion, although unfortunately high overall intensities are usually required.

To further confuse the issue, sometimes the term "electronic nonlinearity" is used to emphasize that a nonlinearity works directly with the electronic transitions (e.g. as in absorption saturation of an atomic transition with real populations or as in second harmonic generation with virtual transitions), rather than through, say, some thermal effect. "All-optical" is sometimes used to mean "intrinsic" as defined here, but there is some semantic confusion. In one sense there is no such thing as an all-optical device; every optical interaction involves matter, and every optical interaction that interests us here involves electrons. Another reasonable use of the term is to describe systems in which there is essentially no electrical communication of information outside of the device, so that from the point of view of the system in which the device is embedded, the device is "all-optical".

Devices involving the mixing of waves of substantially different frequencies are possible, but unfortunately the nonlinearities are usually weak because they are usually virtual effects. Consequently, such devices have received less attention because of the severe optical power constraints. Also virtual transition devices can have severe problems with coherent back-coupling because of the inherent reversibility of the processes. The majority of devices can therefore be described as either absorptive or refractive (or both in some cases), since in the absence of wave mixing we can describe the effect of the medium on the beam in terms of the properties of the dielectric constant at the particular frequency of interest. Here we mean absorptive to include gain effects since gain is negative absorption. If we imagine a light beam making a single pass through an absorptive medium of transmission $\exp(-\alpha l)$, then it is clear that in order to make a substantial change (e.g. a factor of e or more) in the transmission of the light through the medium of thickness l (i.e. to "switch" it), we must change the number of absorption lengths (αl) by one or more, i.e., $\Delta \alpha l \geq 1$. The energy required to do this to the material will give us a measure of the minimum switching energy of the device. For a single pass refractive device such as a Mach-Zehnder interferometer, to turn the device from totally on (constructive interference) to totally off (destructive interference) requires a change in optical path of half a wavelength, i.e., $\Delta n l = \lambda/2$, and similarly the minimum switching energy is the energy required to effect this change in the material. Interestingly, the changes in the magnitude of the (complex) dielectric constant ε in each of these two cases are rather similar. (Strictly, the magnitude of the change that in the real part of $\sqrt{\varepsilon}$ would give $\lambda/2$ path change would instead alter the absorption of the material by 2π absorption lengths if made in the imaginary part of $\sqrt{\varepsilon}$.) Incidentally, although it is not immediately obvious, may other kinds of refractive device also require the same $\sim\lambda/2$ path change for switching, even although they are not obviously interferometric; this is generally true, for example, for self-focusing and self-defocusing devices, for devices in which a beam is deflected, and for devices in which a waveguide is changed from guiding to non-guiding.

By the use of multiple passes through the same material, we may reduce the required material volume (or the degree of change in the same volume) proportionately, and this is one of the great advantages of cavities, where the volume required reduces in proportion to the cavity finesse F. Cavities can be used both for absorptive and refractive devices. Note that when a cavity is used, the beam still acquires changes in path length or amplitude of the same magnitude as for the single pass device because it makes essentially F passes through the material. For all devices in which linear absorption of power results in the changes in refractive index required to switch the device (as is common in many proposed devices), we must obviously be able to acquire $\lambda/2$ path length change in less than, say, one absorption length in a single pass device so that the device has some reasonable transmission. From the above argument, we can see that this is still the criterion for the material even in the presence of a cavity. In general, for refractive devices, the linear absorption coefficient must be considered in evaluating the usefulness of nonlinear materials. This is not of course a problem in active devices such as laser diodes operated with gain rather than absorption.

4.

SYSTEMS REQUIREMENTS ON DEVICES

This article is not primarily about systems, but, as mentioned in the introduction, devices cannot usefully be considered without looking at them in the context of systems. Many apparently interesting devices are regretably not usable in large logic systems because they fail some basic requirement such as cascadability or are extremely inconvenient to utilize, and some device that is not apparently interesting because its physical performance is unexceptional compared, say, to electronic devices, might be very useful in a system because it would allow us to use the communications ability of optics to construct a better overall system. I will summarize very briefly below some of the considerations imposed by systems on devices.

Digital logic absolutely requires two attributes aside from the ability to perform a basic logic function: (i) cascadability, so that one gate can drive the next, which requires compatibility of input and output format (e.g. voltage in electronic systems, wavelength in optical systems) (ii) fanout, so that arbitrary logic systems can be constructed, and also to provide the gain necessary to overcome loss in the system. At least some of the devices must have a fanout of 2 or larger (i.e. the ability to drive at least two subsequent devices). Somewhere in the system we must also have the option of performing logical inversion. Any Boolean logic function can be constructed from combinations of two-input NOR (or NAND) gates with a fanout of two.

Also desirable for ease of system design are (i) absence of critical biasing, (ii) input-output isolation, (iii) logic level restoration, (iv) flexibility of functionality. The first two are rather obvious, and have been discussed in the previous section. The others require some more discussion. One of the important features of digital logic is that it is infinitely extensible, i.e. we can construct a logic system with an arbitrary number of levels of gates or logic. One reason for this is that the logic levels are "restored" by each gate, i.e. regardless of the precise input levels, as long as they fall within some defined acceptable range, the output logic levels are always essentially the same. In electronics, this primarily relates to voltage levels. In optics it would relate to power levels and also to beam quality, which is always only degraded by the optics of the system and must also be "restored". "Flexibility of functionality" is a more subtle concept. In designing any large and complex system, we must have complexity, i.e. if a system must perform a complex function, it must have involved a lot of choices to design it, because that complexity must be built into the design. The complexity may exist at a number of levels: in the software, in the physical interconnection and layout of the hardware, and in the devices themselves. It is consequently desirable to have choices at the device level as to the precise function the device will perform. Otherwise we must force all the complexity into the rest of the system, making it more difficult to design. In electronics for example, although we could build an arbitrary logic system using only NOR gates, in practice we will not do this, because the availability of devices with other functions (e.g. flip-flops, AND gates, etc.) allows us to make a simpler system overall. Very few classes of optical devices currently have much functional flexibility. Devices that under different biasing conditions can perform different functions do not necessarily solve this, because they simply transfer the complexity to the design of the biasing system; ideally, we would like to have a choice of devices that, under the *same* biasing conditions could perform different functions.

To make the system physically reasonable, we also require (i) sufficiently low optical operating energy, because of the finite optical power available, (ii) sufficiently low total operating energy to allow thermal dissipation, (iii) compatibility with relevant optical systems (e.g. planar waveguide devices are not readily suitable for two-dimensional optical processing optics since they are only conveniently available in one-dimensional arrays), (iii) sufficiently fast to allow the whole system to perform the desired task better than other methods. All of these properties relate very directly to the basic physics of the nonlinear processes in use. In discussing energies, it is very important that we consider the energy of the device in the system; we must consider not only the energy required to switch the device, but also the energy required to communicate the results to the next device. Here optics may have a considerable advantage, since it is not constrained by the same communications methods as electronics.

The various features described above have different degrees of importance in different kinds of system. Two extreme kinds of systems that illustrate these differences, and which are both of considerable interest, are (a) large, complex, two-dimensionally parallel processors, and (b) small, simple, fast serial processors. In the large complex processor, all the aspects related to convenience of design are particularly important, as is the possibility of two-dimensional operation, and low switching energies may be especially crucial. Very high speed may not be very important, because the speed of a large system may be constrained by the time taken to communicate from one side of it to the other, and also the limit on speed may be more our ability to provide and sink sufficient power to run the devices. As an extreme illustration, to a run a system with a total throughput of 1 Tb/s per chip with a 1 W laser requires less than 1 pJ system energy per bit (the device switching energy would probably have to be lower than this by a factor of 10 - 100 because of system losses and margins). Relatively simple optics could address 1000 - 10,000 such devices, thereby requiring operating speeds of 1 - 10 ns. Even with only such moderately fast absolute speeds, such a system, if it could be built, would have a level of total information on and off the chip that is very difficult to contemplate electronically. Such a system might be useful in processing tasks, such as telecommunications switching or image processing, that have a high total throughput with relatively simple processing. At the other extreme, we could imagine a very high speed serial task such as time-division multiplexing onto and off of an optical fiber. Here the system might be quite simple in that only a few devices might be required. We could therefore use waveguide devices. Because only a few devices are involved, dissipation of the energy is not problematic, and we can afford to work with slightly less convenient conditions because the system is not otherwise complex. In this particular operation, the total time required to perform the multiplexing, from the time the information enters the multiplexer to the time it leaves (i.e. the "latency") is also not important, and hence we may contemplate the use of very large devices, such as nonlinear optical fiber switches, to perform the task; because the latency is unimportant in this system, we may be able to use the weak but very fast virtual transition nonlinearities with a very long interaction length in the fiber. Overall operating power might also not be such a problem because we are only using a few devices, but clearly speed would be very important. It is debatable whether we currently have the optical devices to address either of these example cases, but the point is that different systems have very different device needs.

Finally, the system must be capable of performing the desired task sufficiently cheaply. One important aspect here is that device cost and device complexity are not closely related except within a given technology. A complex device within one technology may be cheaper than a simple device in another (e.g. a silicon chip compared to a laser diode). In considering cost, again we must not consider the device in isolation from the system; we must also consider, for example, the cost of interconnecting the devices, and this again may turn out to be one of the major advantages of optics. A crucially important aspect of determining cost is "re-use" of existing technologies. Unless and until optics establishes a serious role in switching and processing, it will not be possible to justify developing a completely new technology for such applications, hence technology re-use is vital. It is no accident that there is considerable effort to use semiconductors as the optical materials because their technology is already well developed for other reasons. Similarly, we may try to make use of laser diodes developed for communications or other applications, or optical coating technology, or semiconductor lithographic or layered growth techniques for the fabrication of complex devices.

5.
CONCLUSIONS

In summary, we can see that there is a very large variety of types of nonlinear optical processes, and also a large selection of different devices that have been proposed. At the time of writing, we are just beginning to see the appearance of devices that may be seriously usable in systems. An important conclusion of this discussion is that no one parameter of a device can be used as a reliable guide to its usefulness. A device must be considered in the contexts of the system in which it will be used and the technologies that will be used to make it and the rest of the system.

Many of the advantages of optics come at the system level (e.g. in energy, speed and globality of interconnections), not at the device level, and consequently we should not always be looking simply for physical advantages (e.g. lower energy or higher speed) in the devices themselves. It is however important that we have devices that are good enough to make it worthwhile to contemplate practical systems. Some of the devices that are emerging now are seriously attractive candidates because they address many of the issues discussed above for "useful" digital logic devices. If they are successful, we can expect a strong pull for evolutionary improvement of these devices and we will see increasing technological resources devoted to optical systems in general that will be of much wider benefit both in the applications we can foresee and those we cannot.

D. A. B. MILLER

BIBLIOGRAPHY

Because of the very large number of topics addressed here, I have not attempted to give a reference list. Instead, I will suggest some publications for further reading.

For a thorough discussion of "classical" nonlinear optics, see
Y. R. Shen, "The Principles of Nonlinear Optics" (Wiley, New York, 1984)

For recent discussions of semiconductor nonlinearities and devices, including quantum wells and self-electro-optic-effect devices, laser diode switching devices and other semiconductor systems, see the many chapters in
H. Haug (ed.), "Optical Nonlinearities and Instabilities in Semiconductors" (Academic, New York, 1988)

For a summary of work in optical bistability, see
H. M. Gibbs, "Optical Bistability: Controlling Light with Light" (Academic, New York, 1985)

For critical discussions of the requirements on logic devices from the perspective of electronics, see
R. W. Keyes, Science **230**, 138-44 (1985); Opt. Acta **32**, 525-35 (1985); Int. J. Theor. Phys. **21**, 263-73 (1982)

QUANTUM WELL ELECTROABSORPTIVE DEVICES:
PHYSICS AND APPLICATIONS

D. A. B. Miller

AT&T Bell Laboratories, Holmdel, NJ 07733, USA

1.

INTRODUCTION

Quantum wells are one of the more promising developments of recent years for applications in optical devices. They offer low-energy optical mechanisms compatible with laser diode light sources and semiconductor electronic devices, and are particularly well suited to integration with electronics and for fabrication of two-dimensionally parallel arrays compatible with the opportunities of free-space optics.

In this short set of lecture notes, I will give a brief summary of quantum wells, their electroabsorptive properties, and their possible applications in optical switching and computing. Comprehensive reviews of nearly all of this material already exist in the literature [1-4]. For the most part, I will not make specific references to the literature except when the work is too new to be included in these reviews.

I will start by introducing layered semiconductor structures and quantum wells (section 2), and proceed to summarize electroabsorption in semiconductors in general and in quantum wells, introducing in particular the quantum-confined Stark effect (QCSE), which is the mechanism on which all the devices discussed here are based (section 3). In section 4, I will discuss optical modulators, and in section 5 I will summarize the self-electro optic-effect devices (SEEDs), which are the ones most interesting for optical switching (in the sense of light switching light).

2.

LAYERED SEMICONDUCTOR STRUCTURES

The modern techniques of layered semiconductor growth, such as molecular beam epitaxy (MBE) or organo-metallic vapor phase epitaxy (OMVPE, also called metal-organic chemical vapor deposition (MOCVD)), allow us to grow crystalline semiconductor layers with thickness control down to single atomic layers. Within limits related to the mismatch of the lattice constants of the bulk materials, we can grow layers of different semiconductors one on top of the other. We can choose to dope different layers as we wish, and can obtain very abrupt interfaces between materials and between doping levels.

Such growth techniques can be applied to make many different kinds of devices (e.g. sophisticated transistors, laser diodes, optical detectors, optical waveguides, and quantum well devices as discussed here), and have the potential for integrating several different types of these devices together. They also allow the study of various novel physical effects seen only in "low-dimensional" systems. Perhaps the most studied material system is GaAs/AlGaAs. These materials have almost identical lattice constants, and essentially arbitrary layered structures can be grown. Most other III-V materials can also be grown, and InGaAs/InAlAs and InGaAs/InP have received particular attention for long-wavelength optical devices. II-VI materials and group IV materials (e.g. Ge/Si) are the subject of active research. Structures involving controlled amounts of strain resulting from lattice mismatch can be grown, giving new properties.

We will discuss here only the optical properties of one particular class of such structures, namely quantum wells. Quantum wells consist of alternating layers of two different semiconductors. The electrons and holes see lower energy in one semiconductor layer than in the other (a potential "well"). If both electrons and holes have lower energy in the same layer, this is called a "Type I" structure, and this is the only one we consider here. (For electrons seeing lower energy in one semiconductor layer and holes in the other, the structure is "Type II"). The layers can be so thin that the electrons and/or holes are "quantum-confined", i.e., they behave like particles in a box in one direction, hence the term "quantum well". In the case of GaAs/AlGaAs, the electrons and holes both see lower potential energies in the GaAs material, which has a lower bandgap energy, with the AlGaAs layers (having a larger bandgap energy) forming the "walls" or barriers. Quantum wells are further restricted in that the barriers have to be sufficiently thick that tunneling between adjacent well layers is weak; the basic physics of quantum wells is essentially that of one well between two barriers, although we often use many such wells, especially to obtain enough optical absorption.

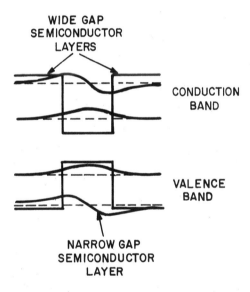

Fig. 1 Wavefunction and potential structure for a simple quantum well, showing two confined states in the conduction and valence bands (not to scale).

Fig. 1 shows some typical wavefunction solutions in a typical potential structure for a quantum well. Strictly, these are so-called "envelope functions" that are a relatively slow modulation in space over the unit cell wavefunctions. In practice, we can usually factor out the effect of the unit cell wavefunctions, which can be viewed as giving the particle an effective mass different from the free electron mass, m_o (effective mass approximation). Then the envelope functions can be calculated by solving Schroedinger's equation treating the particles as having the appropriate effective masses. For the case of simple rectangular structures, this is particularly straightforward, giving sinusiodal solutions within the wells and exponential tails into the barriers for the "bound" states. This is essentially the well-known "particle in a box" solution of the wave equation. In the majority of semiconductors of interest here, the electrons are relatively "light" (e.g. effective mass $m_e \sim 0.07 \, m_o$ for GaAs), and there are two kinds of holes, namely "heavy" holes and "light" holes, with effective masses $m_{hh} \sim 0.4 \, m_o$ and $m_{lh} \sim 0.09$ respectively in GaAs.

3.

ELECTRO-ABSORPTION IN QUANTUM WELLS

Fig. 2 shows a typical absorption spectrum of a GaAs quantum well sample at room temperature. Unlike a normal bulk semiconductor, it shows a series of "steps" in the absorption. These steps result directly from the "particle in a box" quantization discussed above. Without field, only transitions between the same kind of sinusoidal wavefunction in the valence and conduction bands are possible, simply because of the overlap integrals between sinusoidal waves. There is one step corresponding to each such allowed transition. (The actual spectrum is the sum of two such sets of steps because of the two different kinds of holes). It is also clear that there are strong peaks at the edges of the steps. These peaks are exciton absorption peaks. When we absorb a photon to make an interband transition in a semiconductor, we take an electron from the valence band and raise it to the conduction band, but we also leave behind a hole in the valence band, and we must take account of this hole. In fact, a more useful picture is to say we are creating an electron-hole pair. The lowest energy state of such a pair is what is loosely referred to as "the" exciton, which corresponds to the electron and hole orbiting round one another like a hydrogen atom, although with a much larger radius (e.g. $\sim 140 \, \text{Å}$ in bulk GaAs) because of the larger dielectric constant and lower effective masses in the semiconductor. In such a state, the electron and hole have very strong overlap because they are held close together, and, as a result, the optical absorption for the creation of the electron and hole in this state is strong, hence the strong peaks in the absorption. In the case of the quantum well, there is one such exciton peak for every step. Strictly, there are many possible exciton states corresponding to the various excited states of the hydrogen atom, and these have to be included in any complete description of the absorption, although the lowest state is usually the only one to show a strong feature.

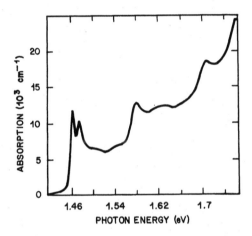

Fig. 2 Absorption spectrum of a GaAs quantum well sample at room temperature for ~ 100Å quantum wells.

The exciton peaks in quantum wells are particularly strong because the excitons are squeezed by the walls of the wells to be even smaller, making the electron and hole even closer, hence giving even stronger absorption. This smaller exciton also orbits faster in the classical sense. In the case of a typical bulk exciton at room temperature, it cannot normally complete a classical orbit before being destroyed by collision with an optical phonon, and hence the exciton line becomes very broad by the uncertainty principle and the exciton cannot really be considered as a stable particle. The faster orbit of the quantum well exciton allows several orbits before such destruction, hence giving still a well-defined peak.

The simplest and best-known electroabsorption mechanism in bulk semiconductors is the Franz-Keldysh effect, which is calculated neglecting any excitonic effects. The effect of applying an electric field can be viewed as tilting the conduction and valence bands. Thus, loosely speaking, an electron can be lifted by a photon with energy less than the bandgap energy to a point below the conduction band from which it may tunnel laterally into the conduction band. This gives a weak absorption tail extending below the band gap energy.

When we include excitons in bulk semiconductors, we must also note that these excitons are very readily ionized by applied electric fields. This gives a broadening of the exciton line even for quite modest fields that is a consequence of the uncertainty principle since the lifetime is reduced by the field ionization of the exciton. When excitonic absorption lines are clearly present, this behavior actually dominates the electroabsorption at least for low fields. Consequently therefore in bulk semiconductors the dominant effect of applying electric field is to broaden the band edge, including any excitonic feature, and generate an absorption tail extending below the bandgap energy.

In quantum wells, for electric fields applied in the plane of the quantum well layers, exactly the same phenomena occur as occur in the bulk semiconductors, in that the exciton absorption peak is strongly broadened with the field and a weak absorption tail appears below the bandgap energy. If, however, we apply the field in the direction perpendicular to the quantum well layers, we see quite different behavior. As we apply the electric field, the exciton absorption resonances remain clear and well resolved, and can be shifted substantially to lower energies with increasing field. The reason for this is that the field ionization of the exciton is essentially prevented because the electron and hole are pulled only to opposite sides of the same well and cannot go any further. In this situation, they are still capable of executing orbits round about one another even though these orbits are somewhat displaced, and hence we still have a well-defined particle that may live for many classical orbits. The energy of this particle has however been substantially changed because the electron and hole are pulled apart by field. The dominant contribution to the energy is simply a term of the order of (1/2)**P.E**, **P** being the polarization of the electron and hole in the presence of the field **E**. In general terms this is nothing other than a Stark shift, but in this case the shift may be several times the binding energy of the particle. This is in strong contrast to Stark shifts of unconfined hydrogen atoms or excitons in which the shift of the resonance with field never exceeds about 10% of the binding energy. The mechanism that applies in quantum wells is therefore described as a Stark effect that is dominated by the special properties that result from the quantum confinement of the electron and hole within the quantum well, and hence can be called the quantum-confined Stark effect. To see this same effect with an actual hydrogen atom would require that we confined it between two walls only ~ 1/2 Å apart and that we applied electric fields of the order 10^{10} V/cm. It is of course important in the quantum well case that the quantum well is significantly smaller then the bulk exciton. Otherwise the exciton could be effectively field ionized just by pulling the exciton and the hole far apart to opposite sides of such a wide well. The actual shifts of the exciton absorption lines are relatively easy to calculate because they tend to be dominated by the shifts of the single particle electron and hole states, and hence all we need to solve is the problem of an electron or a hole in a skewed quantum well corresponding to the quantum well potential plus an electric field potential. There is only a small correction from the change in the Coulomb attraction between the electron and hole themselves. The experimental agreement with such theory for the shifts of the exciton absorption peaks is in general very good.

I show a set of spectra in Fig. 3 for the optical absorption edge of the quantum well as electric field is applied. For quantum wells of the order of 100Å in thickness, fields $\sim 10^4$-10^5 V/cm show substantial shifts of the exciton absorption peaks while still retaining their essential sharpness. The small amount of broadening that is seen in the spectra is not due to field ionization, but rather to practical effects associated with minor imperfections in the quantum wells. The slight loss of area under the exciton peaks as they shift is essentially because the electron and hole overlap has been reduced by pulling them apart.

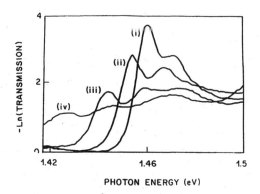

Fig. 3 Absorption spectra for a 94Å *quantum well sample as electric field is applied perpendicular to the layers. (i)* 0 V/cm; *(ii)* 6×10⁴ V/cm; *(iii)* 1.0×10⁵ V/cm; *(iv)* 1.5×10⁵ V/cm.

The changes in absorption coefficients that are induced in these quantum wells in this spectral region are many 1000's cm⁻¹ in absorption coefficient. Thus, with structures only of the order of a micron in thickness we can make significant changes in optical transmission. This change in transmission forms the basis for most of the devices that I will discuss here. There are also associated changes in refractive index, which always occur whenever there is a change in absorption spectra, and these may also be of some interest for optical devices.

4.

OPTICAL MODULATORS

The necessary fields to make optical modulators (and the other devices that we will discuss below) can most conveniently be applied by putting the quantum wells within the intrinsic region of a reverse biased p-i-n diode. For an intrinsic region of thickness 1μm, for example, consisting perhaps of 50 or more quantum wells, applying a reverse bias of, say, 10 volts will apply a field of 10⁵ V/cm, exactly the magnitude required for large optical modulation. With these systems such a structure can be grown, for example, on a GaAs substrate in a MBE machine, and will typically utilize transparent AlGaAs p and n layers. In using such a device, in this simple structure at least, it is necessary to remove the GaAs substrate, which is opaque at the operating wavelength.

It is worth emphasizing here one of the major physical features of this electro-optical effect that makes it particularly attractive from a physical point of view for optical devices. This is that operating energy per unit optical area can be extremely low. The energy required to change the optical properties of a one micron thick layer of this material is of the order of a few $fJ/\mu m^2$, since this energy is essentially CV^2 where C is the capacitance of diode structure and V is the operating voltage. This energy density is comparable to that of a good electronic device, although it is not quite as low as the very best electronic devices because the operating voltage here is somewhat higher.

Such quantum confined Stark effect devices have various other attributes that are also important for practical devices. They will probably operate very fast; to date, this effect has operated as fast as we have been able to apply electric field to it, and the physical limit is expected to be somewhere below a picosecond. This effect is also extremely well suited for the fabrication of two-dimensional, parallel arrays of devices, since we can obtain sufficient change in transmission with only microns of thickness of material. Other semiconductor electro-optical effects do not in general enable us to do this. These devices are also compatible with laser diodes and electronics in fabrication, in operating wavelengths, in optical powers, in electric powers and voltages, and in materials, and are therefore very promising for integrated optical and opto-electronic systems.

Many different optical modulators have been proposed and demonstrated based on the quantum-confined Stark effect. In the GaAs/GaAlAs system in particular many different structures have been tried. These modulators can also operate quite effectively in the waveguide mode with the light propagating in the plane of the quantum wells; this configuration is particularly interesting for the possibilities of integrating laser diodes and modulators within a chip. Other variations on the basic modulator structure include modulators in which there is a dielectric stack mirror, made of GaAlAs and AlAs quarter wave layers, grown directly on the substrate before the growth of the quantum well diode itself. This enables us to avoid removing the GaAs substrate; the light incident from the top of the structure passes through the quantum wells, reflects off the mirror, and passes back through the quantum wells to re-emerge, modulated, from the top surface. This has the additional advantage of offering two passes through the quantum well material, hence improving the contrast ratio of the modulator further. Another variation in modulator design is to change the actual shape of the quantum wells themselves. Although the majority of devices have used simple (rectangular) quantum wells it is possible to use much more complex shapes if we wish. The theory is readily extended to such structures. One example structure is where we replace the simple quantum well with a pair of well layers separated with a very thin barrier - a coupled-well system. This has different behavior from the simple rectangular well but can show a larger reduction in the overlap integral of electrons and holes with applied field, and hence can give enhanced change in absorption from that particular mechanism. In general such different quantum well shapes offer us a way of making engineering improvements to quantum well modulators, although they probably do not offer order of magnitude advantages, at least as so far as we understand them at the moment. The usefulness of such sophisticated structures depends on the particular application in which we are interested. The quantum-confined Stark effect in such diodes can also be used to make tunable photodetectors; the p-i-n structure is of course ideally suited for photodetection since essentially all electron-hole pairs generated in the intrinsic region will result in photocurrent being collected in the reverse-biased device.

Although the majority of the work in such modulators has been performed in GaAs/GaAlAs structures, other material systems have been investigated with, overall, considerable success and promise for devices operating at other wavelength regions. Most investigated have been those materials which have the potential to operate in the longer wavelength regions, such as 1.3 μm and 1.5 μm, that are of most interest for very long distance optical fiber communications. Examples of such systems are InGaAs/InAlAs, GaSb/AlGaSb, and InGaAs/InP. All of these show sufficient electroabsorption with particularly clear shifts being seen in the GaSb/AlGaSb and InGaAs/InP systems. Electroabsorption has also been clearly seen in the InGaAs/GaAs system, which is particularly interesting as it is a strained-layer system. There is therefore considerable promise for extending these effects into many materials systems and hence many operating wavelengths.

Quantum-confined Stark effect absorption modulators have various features in their favor compared to other modulators. First of all they can operate at low voltage, particularly in the waveguide devices where operation with a volt or less is possible. They can also have very low total volume (for example, cubic microns in principle) which is therefore promising for highly integrated devices. I have already mentioned their low energy density capability. They also have the potential to be very-low-chirp modulators (i.e. there is very little frequency sweep as the amplitude is changed), and high speed operation has been demonstrated, at this time up to about 5.5 GHz. Other attractive features include the absence of any need for velocity matching; this is a common problem in modulators in which the overall structure is large and account has to be taken of the different propagation velocities of light through the structure and of the electric field pulse which is to modulate the light. Finally it must be emphasized again that such structures relying on the quantum-confined Stark effect have an almost unique ability to make light modulation for propagation perpendicular to the surface of arrays of devices.

Among the drawbacks of such modulators is the difficulty of taking full advantage of their small size because of stray capacitances introduced in mounting non-integrated devices. They also have a relatively narrow wavelength region of operation (e.g. of the order of a few nanometers for optimum performance). It is also clear that some temperature stabilization will be required to maintain optimum performance. Such stabilization would need to be within, say 5K, (corresponding to the approximately 2Å/K shift of the band edge). This is clearly an engineering constraint but it is by no means a major technical problem. Modulators may also saturate at high intensities (for example $\sim 10 \text{kW/cm}^2$) depending to some degree on the time taken to sweep carriers out of the structure). Another obvious constraint is that these are absorption modulators, and there will always be a tradeoff between the contrast ratio of modulation and the amount of background absorption or loss that will be involved with the modulator. For devices in which the light is propagating perpendicular to the layers, it is not easy to make very high contrast modulators, although of the order of

8:1 contrast has been demonstrated for a reflection devices with the built in-mirrors. For the case of waveguide modulators very high contrast can be made (e.g. 30:1) although again there is some tradeoff involving background absorption and the ultimate performance of the device.

As to refractive-index-based modulators, these devices are under active research at the moment. It is possible to make large changes of refractive index in the quantum wells resulting from the quantum confined Stark effect. The largest index changes occur in those regions of strong absorption (e.g. near the exciton absorption peaks), and hence cannot be utilized efficiently for a device. To achieve the necessary criterion of obtaining half a wavelength path length change in less than one absorption length, it is necessary to operate at wavelengths reasonable longer than the band edge, and hence some tradeoff in performance is again necessary in order to make the background absorption low enough. It does appear, however, that such quantum well refractive modulators may offer significant performance advantages over some other systems.

5.

SELF ELECTRO-OPTIC EFFECT DEVICES

The devices discussed so far have been devices in which the optical transmission is controlled by applying an electrical voltage. Clearly to be able to use this same physical effect (the quantum-confined Stark effect) for devices in which light switches or controls light, we must somehow make the system sensitive to light inputs. The way of doing this is simply to combine the quantum well modulator with some form of light detection, so that light shining on the detector causes a change in voltage across the modulator and hence creates a device with both an optical control input and an optical output. This is the principle of the self electro-optic effect device (SEED). I have already mentioned above that the optical modulators themselves offer very low energies of operation. To take advantage of this to make optically-controlled optical devices with similarly low energies, it is very important to integrate the detector and the modulator so that we do not incur large stray capacitances or other parasitics in the system. If and only if we manage to achieve such an efficient integration will we be able to make devices that can take full advantage of the low operating energy of the quantum-confined Stark effect in optical switching applications in which light controls light. If we are able to achieve such integration however, we will be able to make optical switching devices with extremely low operating energies. These energies are so low that we will be able to achieve interesting devices even without having to use resonant cavities.

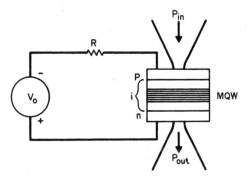

Fig. 4 Resistor-biased bistable SEED circuit.

A large number of such self electro-optic effect devices have been proposed, either in integrated or nonintegrated form, and several have been demonstrated. The simplest of all of these devices is the resistor-biased bistable SEED. This is shown schematically in Fig. 4. In this device the same quantum well diode is used both as the detector and the modulator. For bistable operation we operate at a wavelength approximately coincident with the position of the exciton peak at 0 V bias. In operation, before we shine any light into the device, all of the supply voltage in the circuit shown in Fig. 4 appears across the quantum well diode as a reverse bias. Hence the optical absorption of the diode is relatively low. As we start to shine light into the diode we start to generate a photocurrent, which gives us some voltage drop across the resistor R. This voltage drop means that there is less bias voltage across the quantum well diode, and hence we have a larger absorption within the diode. Consequently, the photocurrent increases somewhat more, giving a yet larger voltage drop across the resistor, and hence yet smaller voltage across the quantum wells, resulting in yet larger absorption in the quantum well diode, and so on. Hence we have a positive feedback mechanism, and this mechanism can become so strong that the device switches into a state of high absorption. This is at the basis of the bistability of this simple system, which is an example of optical bistability from increasing absorption. Such devices are easy to demonstrate, in discrete configurations at least, and the switching power and speed can be chosen over a wide range by choice of the resistor in the circuit. For example, for large discrete devices of dimensions on the order of 100μm, switching speeds of a few 10's of nanoseconds would be typical with switching powers of the order of mW's. This can be scaled reciprocally to much lower switching powers, for example much less than a nW, with a proportional increase in the switching time. In such a simple system, the switching time is essentially the resistive-capacitive time constant of the device in the circuit. Devices with smaller area will switch faster for a given power because their capacitance is lower, hence the critical importance of integrating such devices so that they can be made small without being dominated by parasitic capacitances.

Many other variations on simple devices like this are possible. If instead of choosing to work at the peak of the exciton absorption of zero field, we chose instead to work at a somewhat longer wavelength where the optical absorption will increase as we increase the reverse bias, then instead of getting positive feedback (which gave rise to the bistability) we will obtain negative feedback. If instead of using a resistor and a voltage supply as the external circuit, we use a current source, then this negative feedback mechanism works in such a way that the device adjusts its own voltage so that it absorbs an amount of power proportional to the current being supplied by the current source. Hence we can obtain what is called a self-linearized modulator, in which the optical power substracted from the beam varies exactly linearly with the control current through the device. Such a device can also be used with a constant current input, in which case over a given range of input powers it substracts a constant power from the output, hence performing the function of optical "level-shifting".

The bistability in very simple circuits can be viewed also as being a consequence of negative differential resistance. When we have chosen to operate at the wavelength near the zero field exciton absorption peak as we shine light onto the diode, the photocurrent generated with a constant light power shining on the device can actually decrease as we increase the reverse bias voltage because the absorption of the diode is decreasing. Therefore we have a situation of decreasing current for increasing voltage, hence the negative differential resistance. As a consequence it is possible to make electrically bistable devices out of these systems as well with constant light being shone on the devices. Another consequence, however, is that we make an oscillator. It is well known that negative differential devices operated in conjunction for example with a "tank" circuit, i.e. an inductor-capacitor resonator, can make simple oscillator systems. In the present case, therefore, if we make a circuit with a reverse bias supply and a series inductor (utilizing for example also the intrinsic capacitance of the diode itself), with a constant light power shining on the device we can obtain optoelectronic oscillation, that is, the electrical voltage across the diode oscillates essentially sinusidally in time and also the transmitted optical power shows an oscillation at the same frequency.

Some of these simple systems have recently been combined to make a proof-of-principle demonstration of an all-optical repeater[5]. In this system an incoming bit stream on an optical beam is partly injected together with a power beam from a cw laser diode into a SEED structure operating as an oscillator, with the natural frequency of the oscillator chosen to be near that of the incoming bit stream's clock frequency. Even although the fraction of the incoming bit stream that is injected optically into the oscillator is small and its absolute power is low, it is found that this bit stream can lock the local oscillator to the clock frequency of the bit stream. Thus in such a system we have performed "optical clock recovery". This optical clock signal is now injected together with another portion of the power of the incoming optical bit stream into a decision element, which may be either a SEED optically bistable device or a SEED level shifter, functioning as an AND gate. The output of this latter decision device becomes the retimed, regenerated and amplified version of the incoming bit stream, hence performing in principle the function of optical regeneration. The actual performance of the particular system demonstrated so far is still modest because highly non-optimized and non-integrated devices were used, but with integrated systems and better designed devices, respectable performance may be achievable.

The bistable SEEDs described so far have involved resistive loads. We can however replace the resistor with a current source and achieve improved bistability. Furthermore, this kind of source is particularly well suited for integration. One simple way of making a current source is to reverse-bias another photodiode. The current through such a photodiode will, over a large range of the reverse bias voltages, be essentially independent of voltage and depend instead only on the amount of light shining on the photodiode. The first integrated SEED devices were made using this principle. First a quantum well p-i-n diode was grown on the substrate. Then, after an internal contact, a conventional AlGaAs p-i-n photodiode was grown. This gives two photodiodes electrically in series so that the conventional photodiode can serve as a load for a quantum well diode. When this series structure is connected to a constant external electrical bias supply we can obtain optical bistability seen in an infrared beam, which passes unaffected through the AlGaAs photodiode and is modulated by the quantum well photodiode, with the threshold for switching controlled by a red light beam that is shone on (and totally absorbed in) the AlGaAs photodiode. In this integrated SEED we can make very small structures with essentially no stray capacitance. The only point at which this stray capacitance is important is at the point in the structure where the voltage changes, which is at the junction between these two photodiodes. That junction however is internal to that device and has no other electrical connection to it. Hence there is essentially no stray capacitance. This device therefore can be scaled in size with proportionate improvements in performance. This particular integrated SEED was successful demonstrated and did display the required scaling behavior both in time and in overall operating energy, although this particular structure was not successfully operated faster than a microsecond because of the limitations on the current density in the internal ohmic contact. However 6 × 6 arrays of small devices were successfully demonstrated[6]. An encouraging aspect of this device's operation was the high uniformity that could be obtained over the various devices.

This integrated SEED device could also be used with constant infrared beams but modulated red beams so that a visible image could be transferred into a transmitted infrared image, either bistably or by self-linearized modulation, depending upon the operating wavelength. Thus a spatial light modulator was successfully demonstrated[6]. Another interesting operating mode of this particular device is what can be referred to as "optical dynamic memory"[6]. If we suppose, with particular red power and particular infrared power shining on the device, that we have bistable operation and the device is in one or other of its bistable states, then we may abruptly interrupt both light beams and the device will retain its state for a comparatively long time (e.g. up to 30 seconds). On turning the light beams back on again the device returns to its former state. The reason for this is that the state is essentially stored as the internal voltage over the quantum well diode. When the lights are both removed, there is nothing to change this voltage other than the leakage currents in the diodes, which are intrinsically low and are also relatively well balanced between the two diodes. When the lights are turned back on again the device latches itself back into its former state provided that voltage has not drifted too far. Hence it performs a function that in electronics is known as "dynamic memory", with the particular feature that there is no need for sense amplifiers because the device is intrinsically bistable. Thus we have a memory that can operate at very low average powers while still being able to switch fast when required by operating at high power.

Another SEED configuration that can be viewed as an extension of the above diode biased integrated SEED, is the so-called "symmetric" SEED.[7] In this case instead of using an AlGaAs photodiode as the load for the quantum well diode, we use another, identical quantum well diode. Hence we have two quantum well diodes in series with each other and with a voltage supply (Fig. 5). As before, the sense of the voltage supply is to reverse bias the quantum well diodes. In this case, we generally operate this device with two infrared beams. One beam shines through one quantum well diode and the other shines through the other quantum well diode. We can explain the operation of this device in a similar way to that used to explain the diode-

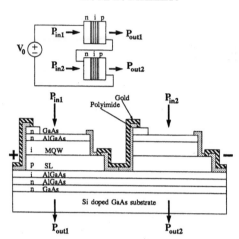

Fig. 5 Symmetric SEED circuit and structure schematic. SL is a short period superlattice used to obtain a high quality structure, and MQW is the multiple quantum well material.

biased SEED device. For example if we shine a constant infrared light beam power into the first diode we will see bistability in the transmission in the second diode as we vary the power incident upon it. We may reverse these roles using the second quantum well diode with a constant light beam power shining on it as the load of the first quantum well diode hence obtaining bistability seen in the other light beam. In fact this device is bistable in the ratio of the two incident infrared light beam powers. To understand why it is the ratio, note that the switching from one state to the other starts when one photocurrent attempts to exceed the other photocurrent. Hence changing both light beam powers by an equal factor does not change the state of the device. Of course, this device has two outputs which are complementary. When one quantum well diode is highly transmitting the other is highly absorbing and *vice versa.* Input/output characteristics as one light beam is varied are shown in Fig. 6.

The operation with the ratio of two light beams has some important and non-trivial consequences. One immediate consequence is that, if we derive both light beams from the same light source, then fluctuations in that light source power do not cause the device to switch. Another consequence is that we may turn down the power of the light source to both devices to a low level, and then switch the device from one state to the other with a small additional power on one of the diodes. Then we may increase

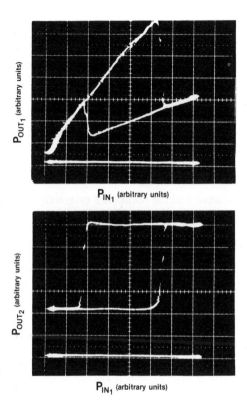

Fig. 6 Input/output characteristics of a symmetric SEED as one input power is varied.

the power from the light source to a much larger level, the state of the device will be retained, and hence we may read out the new state of the device at a high power level. Consequently we can achieve power gain; we only used a small power to change the state of the device but we have caused a large change in the transmitted power when we read the state out at high power. This phenomenon can be called time-sequential gain. This gain can be very large if we wish; in fact with low leakage SEED devices we could obtain gain of many thousands or even millions, although the gain-bandwidth product would remain constant since high gains mean that the device is being switched at very low powers and hence will take a longer time to switch. Another important attribute of this device is that it also has relatively good input/output oscillation. When the device is sensitive to inputs (i.e., when it is being operated at low power), there is relatively little output from the device. However, when it is being operated at high power, so that it has a large output, it is relatively insensitive to inputs. Hence the gain of the device does not amplify small fluctations fed back into the output of the device from the output beams themselves. Thus this device is, in effect, a three-terminal device; it shows gain without critical biasing, it is insensitive to power supply fluctations, and shows relatively good input/output isolation. It is interesting to note that in this device the inputs and the outputs are applied or can be applied at the same physical points of the device. One might ask how then it can be a three-terminal device. The answer is that input and the output are essentially orthogonal in *time*, and so in the most general sense of the definition of a three-terminal device this device is indeed such a device. Although it operates in a slightly unusual fashion for a three-terminal device, this three terminal nature does indeed make it much easier to contemplate making systems with such a device. The unusual requirement of varying the power supply in order to obtain the gain is quite compatible with clocking schemes that would be required in use of such devices in systems. The use of a pair of light beams to carry the information from one device to another also has several attractive aspects. In this case the state of the beams is conveyed by whether or not one beam is larger than the other, and not by the absolute power in either beam. Thus the logic state is not affected by slight attenuations of both light beams together.

Quantum well SEED devices can also be extended to include more sophisticated circuits between detector and modulator. An obvious example is to attempt to include electronic transistors to provide gain or increased functionality in the device. Various schemes involving the bipolar transistors and modulators have been proposed although these have not yet successfully been integrated. One system involving field effect transistors and modulators has been demonstrated in a partially integrated form.[8] This field-effect transistor SEED (F-SEED) consists of field-effect transistors fabricated in the top layer of the modulator structure itself. The fabrication of the field effect transistors is essentially identical to standard metal-semiconductor field effect transistor (MESFET) fabrication techniques. However with this integration we find that there is a modulator available underneath each and every transistor if we choose to use it, since the drain of the transistor is automatically connected to the top layer of the modulator. Hence, as the drain voltage alters in the normal operation of the field-effect transistor circuit, this voltage can be used to modulate a light beam passing beside the drain of the field effect transistor. We can also use other portions of the modulator structure as photodetectors, and hence we can envisage circuits with optical inputs connected to some transistor circuitry to provide enhanced functionality while at the same time being able to extract optical outputs from the same system. Hence we have a way of making intelligent, functional blocks of electronics with optical interconnections. The simplest of such circuits has been demonstrated, with the field-effect transistor, modulator and detector integrated (only the other circuit components, such as resistors, were not integrated in this demonstration).[8] This circuit involved only a single field effect transistor with a single photodiode and modulator, but could be operated as an optical amplifier showing a differential gain of 25 between the variations in input power to the photodiode and the variations in transmitted output power from the modulator. When looked at from the point of view of electronics, this method offers a potentially low energy technique for extracting optical information from electronic circuits, since the optical output "pads" in this system need only be of micron dimensions as far as the optics are concerned, hence comparing very favorably with typically hundred micron dimensions of electronic output pads.

There are limits to the performance of SEED devices. I have already discussed above the basic electrical energy limitations; in devices without electrical gain, the optical energy requirements are that the optically created charge be able to charge or discharge the device capacitance. The resulting optical energy requirements are rather similar to the electrical energy requirements, i.e., in a range of 1 to 20 $fJ/\mu m^2$ micron depending on the voltage and device type. Speed limitations on SEED devices will most likely be due to limits in the photodetection process. The quantum-confined Stark effect itself can probably operate faster than 1 ps although at present measurements of the speed have been limited by external circuit capacitances to of the order of 100 ps. Photodetection on the other hand will be intrinsically limited by carrier emission times from the wells; transport data show, in preliminary experiments, times of 30 to 100 ps for this process. SEEDs have been successfully tested down to switching times of the order of 1 ns. There will also be limits in both modulators and SEEDs to the maximum power that can be passed through the modulator parts of these devices. One limiting mechanism is that, if the optically-created carrier density becomes too high, the exciton absorption may be saturated. This process is much less severe in quantum wells with field applied because the fields sweep the carriers out. Saturation intensities in the range of 3 to 10 kW/cm^2 have been observed in GaAs devices. InGaAs/InP devices seem to have more of a problem probably because of another mechanism, namely that it is difficult to extract the holes from the rather deep wells encountered in this material, hence leading to space charge effects that quench the electroabsorption.[9]

6.

CONCLUSIONS

It can be seen that now layered semiconductor growth technology is in a very advanced state, and that we are able to reuse this technology to make novel electronic and optoelectronic devices, of which the quantum well modulators and SEEDs are an important example. The low operating energy of these devices and their compatibility with semiconductor materials and devices and also laser diodes makes them particularly attractive for optical switching applications and integrated optoelectronics in general. A very important aspect of this capability for optoelectronic integration is that we are able with these devices to tailor the functionality of the device to suit the system. At this point we are just starting to see the fruits of this ability. For example the symmetric SEEDs and the field effect transistor SEEDs discussed above are showing that we can use the integration abilities to make devices more suited to practical systems. In general we might hope that this kind of integration will enable us to choose what we want to do in electronics and what we want to do in optics, with a relatively easy interchange between one and the other that does not incur gross inefficiences either in power or in cost at the transition between the two technologies. If this can be achieved it will greatly enhance the abilities of electronics and the usefulness of optics for processing and switching applications.

1. For an introductory tutorial review on quantum well electroabsorption and devices, see D.A.B. Miller, "Electric Field Dependence of Optical Properties of Quantum Well Structures" in "Electro-optic and Photorefractive Materials", ed. P. Günter, (Springer-Verlag, Berlin, 1987), p. 35.

2. For a review of quantum well linear optics and nonlinear effects related to absorption saturation, see D. S. Chemla, D. A. B. Miller and S. Schmitt-Rink, "Nonlinear Optical Properties of Semiconductor Quantum Wells", in "Optical Nonlinearities and Instabilities in Semiconductors", ed. H. Haug, (Academic, New York, 1988).

3. For a review of electroabsorption physics and devices in quantum wells, see D. A. B. Miller, D. S. Chemla and S. Schmitt-Rink, "Electric Field Dependence of Optical Properties of Semiconductor Quantum Wells" in "Optical Nonlinearities and Instabilities in Semiconductors", ed. H. Haug, (Academic, New York, 1988).

4. For a short review of some recent device work, see D. A. B. Miller, Opt. Eng. **26**, 368-372 (1987)

5. C. R. Giles, T. Li, T. H. Wood, C. A. Burrus, and D. A. B. Miller, Electron. Lett. **24**, 848-850 (1988)

6. G. Livescu, D. A. B. Miller, J. E. Henry, A. C. Gossard, and J. H. English, Optics Lett. **13**, 297-299 (1988)

7. A. L. Lentine, H. S. Hinton, D. A. B. Miller, J. E. Henry, J. E. Cunningham, and L. M. F. Chirovsky, Appl. Phys. Lett. 2, 1419-1421 (1988)

8. D. A. B. Miller, M. D. Feuer, T. Y. Chang, S. C. Shunk, J. E. Henry, D. J. Burrows, and D. S. Chemla, Paper TUE1, Conference on Lasers and Electro-optics, Anaheim, April 1988

9. I. Bar-Joseph, G. Sucha, D. A. B. Miller, D. S. Chemla, B. I. Miller, and U. Koren, Appl. Phys. Lett. **52**, 51-53 (1988)

OPTICAL CIRCUITS

S.D. Smith

Department of Physics, Heriot-Watt University,
Riccarton, Edinburgh EH14 4AS, Scotland, UK

1.

SYSTEM REQUIREMENTS ON OPTICAL CIRCUITS AND THEIR COMPONENT LOGIC ELEMENTS

Definitions

The term 'optical circuits' derives from the standard usage in electronics but with the substitution of photons for electrons it does not necessarily have the same implications that the use of the term 'circuit' in its electrical context has. In particular, an optical circuit need not be closed.

Circuits are used for many purposes; in this context we can usefully divide these purposes into two:

(i)   to receive, process, store and transmit information - by implication optically and

(ii)   to condition optical beams for power control (or possibly) material processing roles.

In these lectures we shall be primarily concerned with the first and only with the second in that it impinges upon the first.

Digital Optical Circuits

Optical processing in analogue form is in special cases very powerful but suffers from the familiar disadvantage of analogue electronic information processing that the accumulation of error due to noise and distortion inevitably occurs as such signals are passed through increasing numbers of stages of processing. In the context of computation such error accumulation is fatal. We, therefore, with the requirement of billions of steps of computation, concentrate upon digital optical circuits. Discussion of noise propagation through a cascade of operations has been given, for example, by Sawchuk and Strand[1]. In terms of probability of one or more of the outputs of the system being incorrect for the two cases of analogue and digital processing, this is given respectively by:

$$P_A = 1 - \prod_{i=1}^{N} 2erf(z_i) \rightarrow 1 \text{ as N increases} \tag{1.1}$$

$$z_i = \delta x/\delta i$$

$$P_D = 1 - \prod_{i=1}^{N} 2erf(z_1) \tag{1.2}$$

$$\text{and } 0 \leq erf(z_i) \leq erf(z_1)$$

These relations show that, whereas the analogue system has an accumulation of error, the digital case shows a probability of the ith stage which looks just like that at the first stage and all stages individually have the same error probability. This is another way of saying that errors do not accumulate.

Components

The elements of digital optical circuits will include:
optical logic elements (whether all-optical or optoelectronic in mechanism) capable of receiving and transmitting information through photons, *arrays* of these elements with parallelism contemplated in the range of $10^4 - 10^8$, *interconnecting optics, power supplies* (either optical or electrical or both) capable of deployment in massively parallel arrays, *data storage* devices on *short* or *long-term* scales, *input* and *output* devices, *control* and *clocking* devices.

The major proposition is that information processing and transfer using photons will be able to achieve some functions either better or cheaper than electronic methods.

Physics of Computation

Given the structure outlined above, it is astonishing to me that after more than 20 years of consideration of this subject by the research community so little work has been done on, what I would define as, 'elementary optical circuit principles'. We have the example before us of the growth of digital electronic computers and can hardly be unaware of the role that the circuit sub-systems play. Notwithstanding the elegant theoretical work on optical computer architecture and that upon a wide range of nonlinear optical device development, very little attention seems to have been paid to this intermediate area. This is the subject of these lectures.

To begin this study we need to consider the physics of computation systems and to observe the lessons available to us by analogy with electronic circuits. As indicated above, this does not mean copying all aspects.

The need to operate digitally implies that the sequence of logic computations be made indefinitely extensible without error. This means that the signal that represents the information, and which is stored as energy, must be cascadable indefinitely and must be restored at each step. Within any logic family the stored energy per bit will scale with physical size of the logic elements and the process will be characterised by a definite switching energy. The study of the propagation of signal energy through indefinitely long cascades of restoring logic circuits has been done for electronics (e.g. Mead & Conway[2]) but the results apply to all different families of logic. If one considers a chain of identical inverters, fig. 1, then the transfer characteristic between node 1 and node 2 has a region of valid logic 0, a region of valid logic 1 and an intermediate region in which the slope of the transfer function must be greater than 1, at the point where the input level is equal to the output level.

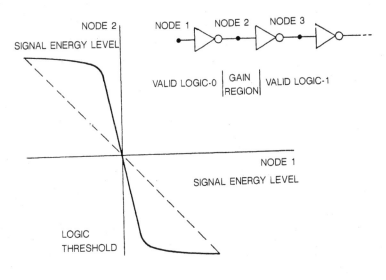

Fig. 1

The need for cascading implies that the first element must supply more energy to the input node of the next stage than has been supplied to its own input node. This cannot, therefore, all come from the switching signal. It must therefore draw power from some power supply separate from the actual signal path. This requirement has important implications

All points so far appear to apply to any logic system and apply equally to electronics and optics. Any proposition that optics can compete with electronics has to be made in the context of at least a million transistors per chip and implications of sizes of 1 μm or less for future devices. As we shall see, however, electronics is constrained in both interconnectivity and signal propagation speed. The different nature of optics may lead to solutions. Any such optical solutions are best applied where they serve system purposes best, preferably in cases where electronics is in difficulty. Propositions about "all-optical general purpose computers" seem premature.

There are some fundamental differences between photons and electrons for the implementation of logic. Let us take the example of an n-MOS FET. In this transistor the voltage controls the current in a simple switch. Electrical current is conceptually a flow of charged particles each interacting with each other. It is always necessary to apply some external agency to maintain a steady current: this is the EMF. By analogy with water flow in a closed circuit, one needs a pump. The EMF is proportional to the rate at which energy is expended and this is always happening in an electrical interconnection (unless it is superconducting). The potential difference providing this source of energy corresponds to its frictional force and to resistive heating by Ohms law. Things are not quite the same in optics. Once a source of light has been generated it propagates effectively without loss in transparent media. As a propagating wave in a transmission line, ExH is analogous to V and H analogous to i with E/H representing an impedance, Z. However, we immediately note that we always detect light as the time average of $E \times H$ which varies as the time average of E^2. This optical 'intensity' (more properly named 'irradiance') is the equivalent of i^2R. An optical intensity logic family therefore lacks the possibility of interconnecting points at the same potential. This manifests itself as an apparent disadvantage if one wishes to construct, say, an n-input AND gate. A true three-port device has two input ports which are independent. Regardless of the magnitude of the signal incident on one port of, say, a two-input NAND gate, the gate output will not switch low unless the other signal is also above the level for valid logic level 1. This is achieved electronically by a circuit such as that shown in fig. 2.

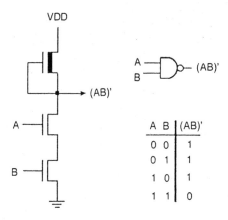

Fig. 2

The inputs to the transistor performing the NAND decision are the outputs from two other transistors - with standardised levels. A discrete optical gate, as we shall see, does not satisfy this condition since the signal intensities simply add. However, the development of optical circuitry for this can ensure that an optical AND gate will function in a proper context. Nevertheless the use of intensity as the only signal quantity will have limitations. Other proposals have been made for use of more than one variable such as polarisation. We discuss purely optical intensity logic in this study.

Comparison with Electronic Logic

It is useful to describe the characteristics of typical electronic logic elements for comparison with our newly developed optical devices - we take a MOS transistor. This is symbolised in Fig. 3 together with its current voltage characteristic.

When there is no charge on the gate the switch is open. Placing electrical charge on the gate controls the number of negative charges that flow from source to drain. The current flow depends upon the driving force V_{ds} and so does transit time

$$\tau = \frac{L}{\text{velocity}} = \frac{L}{\mu E} = \frac{L^2}{\mu V_{ds}}$$ (1.3)

Thus the switching time itself depends on the magnitude of the driving force. We shall see similarity in optical devices. The fastest operation that can be performed is to transfer a signal from the gate of one transistor to the gate of another. The minimum time is τ and to transfer from one to two transistors needs 2τ - with implications for fan-out speeds. The phenomenon of saturation means that as V_{ds} is increased, it is not all available for increasing speeds.

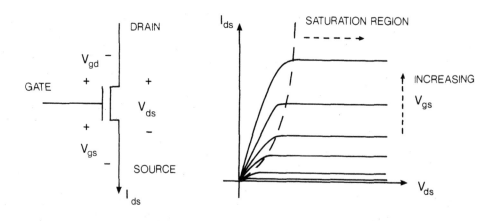

Fig. 3

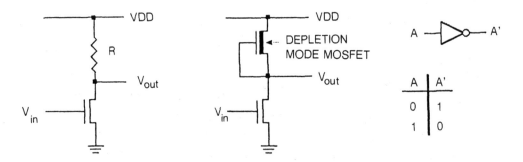

Fig. 4

The simplest, and most important, logic circuit is the basic digital inverter. This extends to NAND and NOR logic gates. Its function is to produce an output that is the complement of its input. A circuit such as in Fig. 4 can achieve this. however, in integrated circuit technology the resitor cannot be adequately achieved and so this is usually implemented with a depletion mode MOSFET, - see Fig. 4, which includes symbolism and the truth table.

In electronic practice there is a change in size of device and of impedence between the pull-up and the pull-down transistors of this device. With the minimum requirement that they must drive another identical device one also encounters inverter delay in driving further devices equal to $f \times \tau$ for the pull-down delay and $Kf \times \tau$ for the pull-up delay where they geometric size ratios of the two transistors are K. The devices find difficulty in driving large capacitance loads leading to trade-offs between distances and delay times. As the devices scale to smaller size the relative delay to the outside world gets larger although the absolute delay gets smaller. We shall see similar features in the very primitive optical circuits which will be demonstrated. It is useful to note that basic NAND and NOR transistor circuits as in Fig. 2 & 5 - include three transistors.

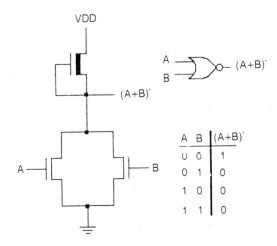

Fig. 5

Stages are often connected with switches known as pass transistors which can be used to minimise inverter delay between stages.

The point of quoting these electronic analogies is to indicate how the simplest logic function has been the recipient of very considerable thought and development of simple circuit procedures. At the present time the properties of a single optical element are often compared with an electronic sub-system circuit. It is not too surprising to find therefore that our devices are at this stage seen to be deficient and my point in emphasising this is to show the need for simple practical and theoretical development at very simple circuit levels. For this purpose it is necessary to have a simple and usable optical logic element that has gain, restoring logic and is capable of driving further elements.

Clocking and Data Transfer

Electronically, data is transferred from 'register' to 'register' through combinational logic - using parallel transistors and inverting logic stages. To do this a two-phase joint overlapping clock system is used to switch and isolate the information as it passes through successive gates. Such processes lead to multiplication of delay times typically 100τ in practice.

If there exist feedback paths, then the forcing function pushing voltage away from a metastable point is proportional to the distance from that point. This piece of physics is common to all bistable technologies and, as we shall see, can affect the switching speeds. In electronics, switching power is energy stored/clock period.

Prise, Streibl and Downs[3] have recently described the simple algebra of the properties of optical logic elements. Their treatment requires the addition of some experimental numbers before conclusions can be reasonably drawn but some of the concepts and equations represent useful statements to develop our discussion further. The physics of electronic devices so far described is in general very similar to the optical devices which we shall be discussing. Bistable devices, for example, can be used either as latches or as gates. The amount of feedback and the initial setting of conditions will affect both the speed and the form of operation. Thus these devices can be used either as thresholding devices or as latches. We shall show that using them as latches combines several of the requirements that require a significant number of electronic components to be built into a circuit. The literature contains some rather uncritical statements implying that 'optically bistable devices are in some way special and more critical'. These conclusions do not usually consider the quantitative physics which is essentially the same for all logic devices e.g. if more output power has to be supplied for cascading than that used to switch the input of the device then a separate power supply is mandatory.

Optical Properties of a Device

A thresholding device is characterised by properties shown in Fig. 6

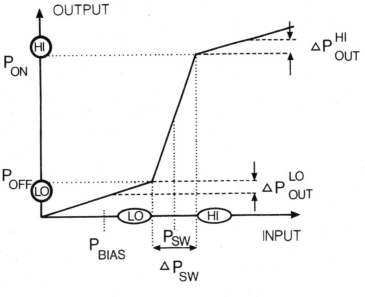

Fig. 6

P_{SW} = the switching power;

the switching window ΔP_{SW} = the difference in input power required to switch the device;

P_{ON} = the output power just after switch on;

P_{OFF} - the output power just before switch-on;

T_{HI} - the differential transmission at the switching point when the device is in the ON-state;

T_{LO} - the differential transmission of the device at the switching point when the device is in the OFF-state.

From this we can define figures of merit characterising the device;

The switching contrast is

$$C_{SW} = \frac{P_{ON} - P_{OFF}}{P_{ON}} \qquad (1.4)$$

The switching transmission is

$$T_{SW} = \frac{P_{ON}}{P_{OFF}} \qquad (1.5)$$

The relative switching window is

$$\sigma_{SW} = \frac{\Delta P_{ON}}{P_{SW}} \qquad (1.6)$$

Note that σ_{SW} can be 0 for a bistable device, but it then switches slowly.

Whether a device is operated in transmission of in reflection is very important; Wherrett [4] pointed out that reflective operation may reduce the switching power, enhance the contrast, reduce problems with spatial hysteresis (deformation of transverse modes of a cavity due to nonlinear index profile), and allow efficient cooling from the back of the devices. The above treatment is easily extended to devices that work in the reflection mode by replacing all the differential transmissions with differential reflections, and by defining the output power as the reflected power. Note also that in some devices the switching transmission can be greater than one. We refer to these devices as having inherent or absolute gain, as opposed to differential gain.

Optical Properties of a System

In addition to the device properties we have to consider the following system properties.

P_{BIAS} - an externally introduced biasing power which can be used to bring the device nearer to its switching point and may act as a separate power supply.

P_{SYS} which we define as the fraction of the total power available from one device to drive other devices. This contains the transmission losses of the optical system used to interconnect different devices.

Variations of the switching power in time and across the array δP_{SW}.

Variations of the bias power in time and across the array δP_{BIAS}.

Variations of the system transmission across the array δT_{SYS}.

P_{OUT}^{LO} and P_{OUT}^{HI} - the output powers of the devices in LO and HI state, respectively.

ΔP_{OUT}^{LO} and ΔP_{OUT}^{HI} - the ranges of output power which are allowed as legal LO and HI, respectively.

All of the inaccuracies intrinsic to the system can be taken into account by defining an effective switching window:

$$\sigma_{EFF} = \sigma_{SW} + \sigma_{SYS} = \sigma_{SW} + \frac{\delta P_{SW}}{P_{SW}} + \frac{\delta P_{BIAS}}{P_{SW}} + \frac{\delta T_{SYS}}{T_{SYS}} \qquad (1.7)$$

Whether we can consider a device as being binary depends on how we operate it. It turns out that the ranges of the output powers LO and HI state depend on the required computational properties of the device, as well as on the system properties. Therefore, for the devices to be binary, the ranges of the output powers have to be small compared with the difference between the absolute output powers of the LO and HI state:

$$\Delta P_{\text{OUT}}^{\text{LO}} \ll P_{\text{ON}} - P_{\text{OFF}} \tag{1.8}$$

$$\Delta P_{\text{OUT}}^{\text{HI}} \ll P_{\text{ON}} - P_{\text{OFF}} \tag{1.9}$$

Analysis of Idealised Binary Devices

If a device switches, its output changes by $P_{\text{OUT}}^{\text{HI}} - P_{\text{OUT}}^{\text{LO}}$. Therefore, on a following device, the input power will change by

$$\frac{T_{\text{SYS}}}{\text{FANOUT}}(P_{\text{OUT}}^{\text{HI}} - P_{\text{OUT}}^{\text{LO}}) \approx \frac{T_{\text{sw}}C_{\text{sw}}}{\text{FANOUT}}T_{\text{SYS}}P_{\text{sw}} \tag{1.10}$$

The first condition for reliable switching is that this input change is bigger than the effective switching window. In terms of fanout, this means

$$\text{FANOUT} < T_{\text{SYS}}\frac{T_{\text{sw}}C_{\text{sw}}}{\sigma_{\text{EFF}}} \tag{1.11}$$

Notice that an ideal bistable device in a system with infinite accuracy can have an infinite fanout. The second condition determines the logical operation that is performed by the device. P_{BIAS} has to be chosen such that:

if there are threshold of 1 or less HI inputs, then the device does not switch;

if there is a threshold or more HI inputs, then the device switches.

This can be written as:

$$P_{\text{sw}} + \frac{\Delta P_{\text{sw}}}{2} - \frac{\text{FANIN}}{\text{FANOUT}}T_{\text{SYS}}P_{\text{OUT}}^{\text{LO}} - \frac{\text{THRESHOLD}}{\text{FANOUT}}T_{\text{SYS}}(P_{\text{OUT}}^{\text{HI}} - P_{\text{OUT}}^{\text{LO}})$$

$$< P_{\text{BIAS}} < P_{\text{sw}} - \frac{\Delta P_{\text{sw}}}{2} - \frac{\text{FANIN}}{\text{FANOUT}}T_{\text{SYS}}P_{\text{OUT}}^{\text{LO}} - \frac{\text{THRESHOLD} - 1}{\text{FANOUT}} \tag{1.12}$$

$$T_{\text{SYS}}(P_{\text{OUT}}^{\text{HI}} - P_{\text{OUT}}^{\text{LO}})$$

We would like to choose the bias power such that is in the middle of the interval given by inequality (1.12):

$$\frac{P_{\text{BIAS}}}{P_{\text{SW}}} \sim 1 - \frac{T_{\text{SYS}}T_{\text{SW}}}{\text{FANOUT}}\left[\text{FANIN}(1 - C_{\text{SW}}) + \left(\text{THRESHOLD} - \frac{1}{2}\right)C_{\text{SW}}\right] \qquad (1.13)$$

This will allow us to calculate the allowable error in the bias beam. To make physical sense, P_{BIAS} must be non-negative. This translates into the following condition for the fanin:

$$\text{FANIN} < \text{FANOUT}\frac{1}{T_{\text{SYS}}T_{\text{SW}}(1 - C_{\text{SW}})} - \left(\text{THRESHOLD} - \frac{1}{2}\right)\frac{C_{\text{SW}}}{1 - C_{\text{SW}}} \qquad (1.14)$$

Equations (1.11) and (1.14) are the basic switching conditions relating the optical and computational properties of a device with the systems properties.

We shall use this nomenclature in discussing the results of some of the early devices which we have to implement optical circuits.

Further System Requirements

The necessity for fanout has been stressed. At this early stage architectural methods are using a minimum of fanout to minimise the practical difficulties of optical technology. In assessing the devices, however, it raises the question of accessing a number of orthogonal modes. One simple method would simply be to image different beams on different parts of our devices. Another way is to design with sufficient flexibility that different angles of incidence can be used. Eventually as the devices are scaled in size this leads to physical limitations. In all cases we consider an array of logic elements and this immediately brings into question the required uniformity and hence fabrication error. Operation of such an array brings with it the need to dissipate any power absorbed. In practice this means that individual element operating power must be less than 1 μW and preferably much less. A wishlist of properties is indicated in Table 1.

TABLE 1

WISH LIST FOR OPTICAL LOGIC PLANE

2-D Array $10^4 - 10^6$ elements	Hold power/element 1 mW - 10 μW
Cycle speed 1 μs - 10 ns	Contrast > 10 : 1
Switch energy 1 nJ - 10 ns	Good throughput: T > 50%
Stability < 1% over hours	Makeable for various wavelengths
Uniformity 1 - 2%	Insensitive to small wavelength change

2.

NONLINEAR OPTICAL LOGIC DEVICES FOR OPTICAL CIRCUITS

A number of purely optical and optical but electronically assisted (optoelectronic) devices have been used for demonstrations of simple optical circuits. These include liquid crystal light valves (LCLV), purely optical nonlinear Fabry-Perot (NLFP) devices and such optoelectronic devices as the self-electro-optic effect devices (SEED). This last class will be described by David Miller in his lecture course. Circuits based on LCLV are described by Stuart Collins elsewhere in these Proceedings. Types of optically bistable systems are shown in Fig. 7.

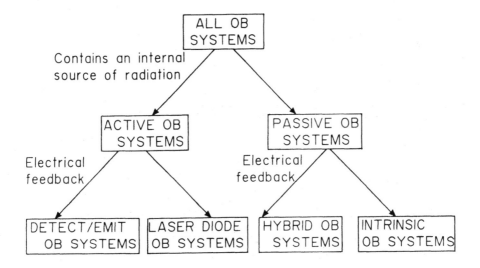

Fig. 7

In this lecture we shall concentrate on one type of intrinsic system, nonlinear Fabry-Perot devices. We restrict ourselves to discussing the device examples where outputs sufficient to drive further, cascadable, stages have been achieved. This excludes a number of reported examples where there are large absorption losses, with consequent small transmission, insufficient temporal stability allowing only transient operation, and multiple wavelength devices.

Nonlinear Fabry-Perot Devices

These are illustrated in Fig. 8 and function through an intensity dependent optical length.

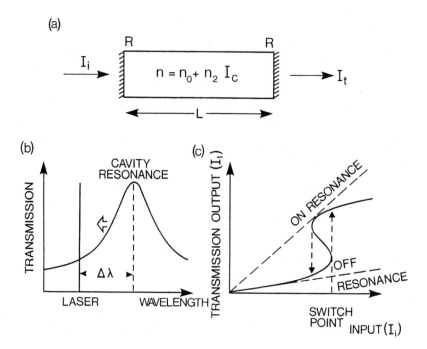

Fig. 8

Imagine such a device illuminated by a laser, detuned by $\Delta\lambda$ from one of the interferometer resonances. As the intensity is increased the optical thickness changes and moves the resonance (say) towards the laser wavelength. The transfer characteristic is now nonlinear. However, the effective field driving the nonlinearity is the internal cavity field I_c. This depends upon the transmission of the cavity and its reflectivity through the relation

$$I_c = I_i T(\lambda)\frac{1+R}{1-R}$$

and hence the phase thickness changes according to

$$\Phi(I_c) = \frac{2\pi}{\lambda}(n_o + n_2 I_c)L \qquad (2.1)$$

and the transmission T through

$$T(\phi) = \frac{T_{\text{MAX}}}{1 + F\sin^2\phi} \qquad (2.2)$$

This is *positive optical feedback* embodying the feature that on-resonance with, say R = 0.9, the circulating internal intensity may be some 20 times greater than off resonance. Consequently, the amount of input intensity required to 'hold' the device on resonance becomes smaller than that to reach resonance. This leads to optical hysteresis as shown in Fig. 9.

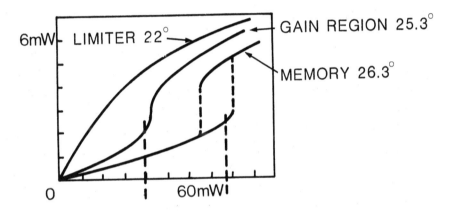

Fig. 9

The actual form of the transfer characteristic can be readily adjusted by setting the initial detuning $\Delta\lambda$. A practical example is shown in Fig. 9 for the case of a nonlinear interference filter (NLIF) where its initial detuning can be set simply by adjusting the angle of incidence between the filter and the incident laser beam. A set of transfer characterestic ranging from a power limiter through a logic element with a gain region to a latchable memory is obtained. The nonlinearity utilised in the above device is the intensity dependent refractive index; nonlinear absorption can also be used. In the refractive case there are at present three potentially usable examples:

(i)   pure optical nonlinearity due to electron motion in semiconductors.

(ii)   electronic effects assisted by the feedback of an injected electrical current.

(iii)   activation of the lattice vibrations of the material by laser heating with consequent modification of refractive index.

S.D.Smith

We will consider (i) and (iii). Conventional wisdom circa 1976 was that electronic optical nonlinearities required large laser powers to be significant [5]. However a discovery in this laboratory in 1976 [6] showed that there was a nonlinear effect that took place with milliwatts rather than megawatts. When the power level of a Gaussian beam from a carbon monoxide laser 5.5 μm incident on a crystal of InSb cooled to 80 K is increased above 30 mW a defocussing effect was observed. This indicated that there was nonlinear refraction which was 10^9 times larger than conventionally expected.

This material, InSb, continues to act as an 'hydrogen atom' for the field. The nonlinearity is so large $n_2 = 1 cm^2 kW^{-2}$ that a change of refractive index of 10^{-2} can be achieved with modest laser powers. Devices only required $\Delta n/n = 10^{-3}$ so that devices operating with less than a milliwatt have been achieved.

The two III-V semiconductors InSb and GaAs were the first to be exploited in 1979. A particular point of interest is the stability. The transfer characteristic was sufficiently stable that the InSb device could be 'set' at any level and remain there all day. Consequently it was easy to use a 'hold and switch' technique in which the hold beam has the required characteristic of providing the 'power supply' such that a small signal beam can delivery a much larger output to the succeeding element (equivalent to power gain). This early result gave some optimism for use in optical circuits [7].

Theoretical explanation of the large nonlinearity was sought and found in a band-filling model [8]. The characteristic n_2 is found with this model to be proportional to τ/E_g^3 where τ is the carrier lifetime and E_g is the energy gap. This result reveals that scaling must take account of both wavelength and speed. Concurrently and independently work, in 1979, at Bell Labs gave similar results on GaAs thin films but with the difference that they had to be illuminated with microsecond duration pulses due to unwanted laser heating in strongly absorbing material ($\alpha \sim 10^{-3} - 10^{-4} cm^{-1}$). Initally the origin of the nonlinearity in this GaAs case was thought to be predominantly due to an exciton effect, negligible in the case of InSb. However, later analysis by the Arizona group [9] showed that the band-filling effect was more significant and consequently one would expect this nonlinear effect to scale as $1/E_g^3$ also. This does not necessarily mean that GaAs devices are at a disadvantage. The scaling of the devices themselves to small size favours the shorter wavelength required (0.8 μm) and tends to compensate for lower nonlinearity. However, stability is equally important.

In the InSb case there are disadvantages of the long wavelength and although the material quality is good ($\alpha = 10^{-1} cm$ the optical working of a single crystal produces some difficulty when constructing a uniform array. As a standard, however, I quote the latest results from InSb for an etalon of thickness 50 μm with a 50μm spot diameter which operates with 650 μ W. Switching times are around 200 ns leading to switching energy for this device of 100 pJ. Scaling to a spot size comparable to the wavelength could in principle reduce this to 1 pJ.

GaAs-based NLFP devices have recently been progressed at Bell Labs [10] with pixel dimension of 2 μm and at CNET [11]. At CNET 4 mW power levels for 10 μm spot diameter and 20 ns switching times have been observed using GaAs/GaAlAs interference filter structures, and millisecond stability obtained by close attention to heat sinking. Switching energies are 100 pJ as for the larger InSb devices but the irradiance needed is $500\mu W\mu m^{-2}$ compared with $0.1\mu W\mu m^{-2}$ for InSb. These operating parameters are beginning to approach those of the Wish List (Table 1) and there is certainly the prospect of further progress.

InSb NLFP's were responsible for some of the first optical binary logic circuits constructed with all-optical intrinsic devices. Using the 'hold and switch' technique the circuit shown in Fig. 10 was constructed in 1982 [12].

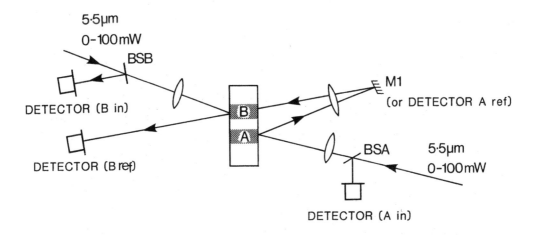

Fig. 10

Opto-thermal Devices

An early report from the Soviet Union [13] encouraged us to investigate the possibilities of using laser heating in an interference filter as a usable nonlinearity for a first generation of optical logic elements for circuit experiments. Over the past 5 years, work has proceeded both at Heriot-Watt University and at OSC, Arizona [14] on this topic. It is possible to create an effective volume of a few μm^3 by both tight focussing of a laser beam and by physical pixellation of the substrate.

Fig. 11 shows that switch power and switch speed scale directly with spot diameter when the spot diameter is very much bigger than the wavelength.

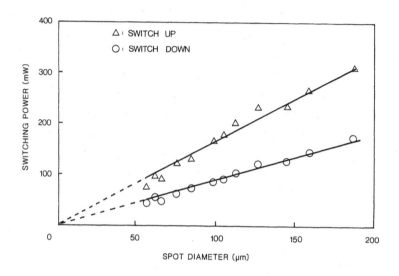

Fig. 11

In our work reasonably stable cw operation with powers of tens of milliwatts was achieved quickly, now improved down to a few milliwatts but stability over hours was not manifest in the early experiments. Recently we have used molecular beam depositon (MBD) ZnSe as the nonlinear material and obtained stability over hours combined with time constants sufficient to cycle a device on a time scale of 100 μs. Very substantial optimisation of multi-layer design and of thermal engineering is possible. In addition uniformity over square centimetres is standard technically in optical thin film coating. Theoretical operating powers of microwatts have been predicted for dimensions of a few microns[16]. The latest approaches and results with the beginnings of thermal engineering and pixellation are given [15]. We have found these devices to be useful building blocks for circuit experiments. The interference filter technology primarily relies upon ZnSe as the active material for which dn/dT is approximately equal to $2 \times 10^4 K^{-1}$.

Obviously, an improved performance could be obtained with better values of dn/dT. The lowest power at which bistable switching is possible can be written

$$P_c = \frac{\lambda}{D} \frac{1}{\frac{dn}{dT}\frac{dT}{dPa}}$$
(2.3)

Lloyd & Wherrett [16], in this laboratory have recently exploited planar-aligned nematic liquid crystals to improve this parameter in a structure shown in Fig. 12.

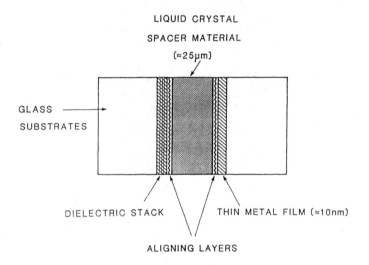

Fig. 12

The liquid crystal 4-cyano-4'-pentylbiphenyl (K15, 5CB or PCB) shows an increase in magnitude of its thermo-optic coefficient as the nematic isotropic transition is approached. Over an ambient range of just 8°C, dn/dT increases by an order of magnitude to a value of $2 \times 10^{-2} K^{-1}$ at 34°C. It was possible to utilise proximity to this transition with a specially constructed temperature control cell allowing choice of temperature to 0.01°C with the result that stable switching was obtained with powers as low as 14 µW for a beam diameter of 25 µm. The reflection and transmission characteristic of the device is shown in Fig. 13.

Both the NLIFs and liquid crystal NLFPs have been operated with diode laser sources. The bistable etalon with absorbed transmission (BEAT [17]) technique is used in which an extra absorbing layer is added without constraining the optimisation of cavity design. Both these devices can be operated over a wide wavelength range since dn/dT does not resonate strongly.

We are therefore currently cautiously optimistic about the prospects for the thermo-optical devices since they scale according to physical laws reasonably well understood and their fabrication technology lends itself to production of uniform massively parallel arrays.

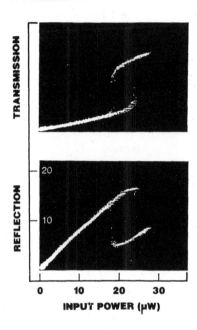

Fig. 13

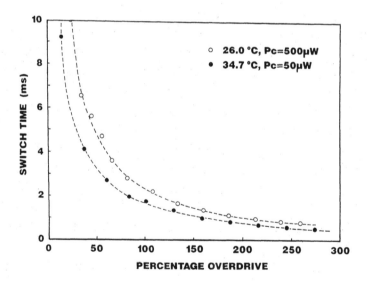

Fig. 14

Adequate stability for circuit experiments is proven. The necessary thermal engineering is likely to be required with device classes (i) and (iii). In any event power levels now seem certain to be adequate: one of the remaining questions is that of speed. Fig. 14 shows the experimental results for switching speed versus - 'overswitch' - the amount by which the incident power exceeds the minimum for the InSb liquid crystal from a recent experiment.

The figure shows that overswitch by a factor of two allows the device to approach its fundamental time constant. Very similar overswitch curves have been obtained for interference filters and InSb. These first experiments give a switch on time of 500 μs for 300% overswitch in the slow liquid crystal case. However, the crucial parameter is the recovery time. This can be arbitrarily improved by increasing the thermal conductivity of the substrate with a consequent increase of operating power. The question is can one obtain a workable trade-off? Table 2 shows some latest results using an interference filter deposited onto polyimide insulating film spun onto a highly thermally conducting sapphire substrate. Although we have thermally modelled this device in some detail, the results have given a surprisingly rapid response with some promise for future development. Similar considerations could be applied to the design of liquid crystal devices. A target in the range 1 - 10 μs therefore does not seem unreachable.

Table. 2

Variation in Device Dynamics with Substrate.

Substrate Description	Critical Sw. Power (mW)	Switch-off (1/e) (μs)	Switch-On (μs) (approx)	Switch Energy = (Crit.Sw.Power) x(Switch-Off)(nJ)
Float Glass	9.0	150	400	1350
Coverslip (30μm)/ Sapphire	10.5	400	1000	4200
Polyimide (1.2μm)/ Sapphire	9.1	13	67	118
Polyimide (2μm)/ Glass	8.2	150	400	1220

Identical filter structure, Spot diameter $\left(\frac{1}{e^2}\right)$10μm, 514 nm

3.

CIRCUITS

As stated in the discussion of the physics of computation, *cascadability* is an absolute requirement i.e. one element must be able to drive at least one other (similar) element. Hence we also have the requirement that the power supply be separate from the switching signal. The nonlinear Fabry-Perot devices, whether operated as thresholding devices or latchable memories which we have described here can, in some cases, operate in a quasi-static 'hold and switch' mode. We can consequently use the 'hold beam' as the 'power supply'. With the degree of stability and contrast already experimentally demonstrated e.g. with the interference filters, sufficient gain for restoring logic operation is obtained for a modest degree of fanout. The hold beams need not be aligned at the same angles as the switch beams. As a consequence a standard hold beam can be applied to successive elements leading to hard limiting and hence restoring logic. Fig. 15 gives an experimental result. Since a new 'optical power supply' is introduced at each stage, there is no limit to the number of circuit stages.

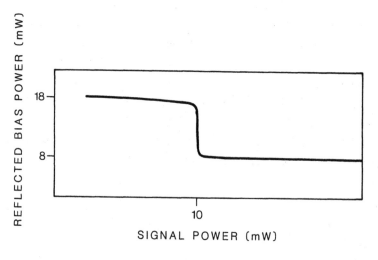

Fig. 15

Data Transfer in an Optical Circuit

Unlike the electronic circuits discussed earlier, the information transfer proceeds through propagation of photons once a logic gate is switched and takes place at the speed of light. As discussed previously, the devices we have used range in speed from a few nanoseconds to almost a millisecond. One must therefore introduce a new delay and a method of isolation of data flow analogous to the two-phase clock of the electronic inter-register transfer commonly practised in conventional digital computers. This procedure, in which electrical control of holding beams is favoured to trap and release information through different stages of the optical circuits, is described in [18] and provides a buffered memory store. Since the clocking times are relatively slow there is every advantage in using a microcomputer to undertake this control. It has become known as 'lock and clock' architecture.

Binary Optical Logic Functions

The nonlinear Fabry-Perot optical element can be used to give a variety of logic functions (Fig. 9). In transmission, for example, with a hold beam set sufficiently below switch point, switching from logic-0 to logic-1 only occurs if two standard inputs are injected. This is the AND function. Using the same device with the same setting in reflection gives the NAND function. Different settings of the holding beam give OR or NOR. It is relatively straightforward to use electronic controls of intensity and clocking (i.e. time control of hold beam) to alter the function of the gates throughout various cycle periods. Before we proceed to discuss the circuits that have been experimentally investigated, I return to the subject of providing adequate power on succeeding circuit elements to switch them quickly. This is hard to achieve with one simple optical logic element. However, Craft [19] has proposed a possible solution in the form of a coupled twin cavity. This device uses two bias beams which have significantly different switching powers for the input and output cavities giving effectively a high gain. The feature of this is that the 'output' cavity is illuminated by a stronger hold beam than the 'input' cavity. In a sense this is rather similar to the construction of the liquid crystal light valve. A much larger output signal is then available for fanout or for speed increase in subsequent stages.

Configuration for Cascading of Elements and for
Restoring Optical Logic

In circuit work at Heriot-Watt University we have made use of ZnSe NLIF technology driven by an argon-ion laser or a GaAs semiconductor diode laser. We have used acousto-optic modulators controlled by a microcomputer to set hold beam and signal intensity levels. The technique for achieving gain, to ensure that the output of one element is sufficient to switch the succeeding one, we have described as 'hold and switch'. By 'holding' the first element as close as possible to switching threshold, an arbitrarily small signal can be made to switch the device - limited only by the stability of the nonlinear element and its 'power supply', the holding laser beam. The change in output can thus be substantially larger than this (extra) switching signal. Thus, if this output is incident on a second identical device held equally close to threshold, it will clearly be enough to switch the second element and thus continue the circuit.

Practical limitations are that the power level of a typical argon-ion laser will fluctuate by 3% and that drift of the NLIF switch power will preclude a closer hold over periods of up to minutes, particularly, when operating predominantly 'on resonance' (where the internal optical field is high). In addition, it is desirable to 'over-switch' by a significant factor to reduce the effects of critical slowing down. We typically hold to 90% of threshold and over-switch by 20% - determined by the output power change of the previous gate. The interconnection optics must preserve this situation for succeeding elements. We use both single and multilens imaging systems to collect and refocus the output signals onto the subsequent gates. Electronics normally separates power supplies from signal channels. We use our optical elements as three-port devices in a somewhat similar manner. 'Hold beams' (power supplies) are supplied to each element separately and the switching signals are introduced off-axis such that they are not collected by the down-line optical aperture (Fig. 15). Thus, provided the input signal is enough to switch the device, the output signal change is standard. This achieves restoring optical logic and, hence, with sufficient independent holding beams, enables the circuit to be extended indefinitely.

'Lock and Clock' Architecture (Buffered Memory Store)

Recirculating configurations are fundamental to many computer architectures. It is therefore essential when investigating the computing potential of any new type of digital gate to demonstrate its ability to pass data, without error, around a loop. We have successfully constructed and operated several such circuits based on bistable NLIF devices.

A 'lock and clock' architecture has been developed to prevent the asynchronous transfer of data round the loop and thus avoid the consequent risk of a new signal appearing at the input to a gate before it has completed its response to the previous one. The 'lock and clock' technique proceeds by stepping the holding beam on each gate between near switch point and zero, thus sequentially enabling and disabling gates. Thus one gate, when enabled, can respond to a transient input, latch (assuming it is held in a bistable region), pass on the datum to the next gate, which is brought from a disabled to enabled state, and finally reset as it is disabled. For a loop circuit containing three gates there is therefore a three-phae clock sequence for the holding beams in which no more than two gates are enabled at any one time. The clocking ensures that the datum is safely passed on to the next gate before each gate is reset. This three-element memory is the minimum required when just on-off holding beam control is used (with three-level holding beam control only two elements are required to store data in a loop).

A Three OR-Gate Circuit with External Input

Figure 16a shows schematically a successfully operated loop circuit based on off-axis addressed bistable NLIFs. Each element, when enabled by a standard holding beam, could be switched by a single input and thus act as an OR gate. The outputs were coupled to the next gates in the circuit. Gate A had two signal inputs, the previous gate output plus an external signal, corresponding to a total fan-in of three beams when the holding beam (power input) is included.

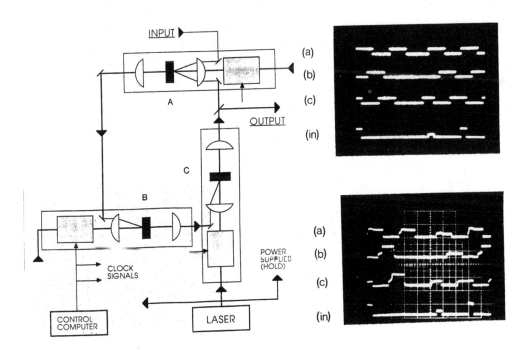

Fig. 16

The three elements were powered by an argon-ion laser computer-controlled acousto-optic modulator (AOM), which set the 'd.c.' levels and imposed the three-phase holding beam clock sequence. The external input signals were determined by a fourth AOM. Figure 16b shows the sequence of gate outputs (observed by detectors after each element). A valid logic-1 input was propagated around the loop in synchronisation with the clock signal, followed, on the next loop cycle, by a signal too small to switch (valid logic-zero), which leaves the three gates in their logic-zero states. (The loop was rest to zero before each full cycle). This transmission-coupled circuit has thus shown that an extensible system of restoring optical logic can be constructed with all-optical bistable devices.

An Inverting OR/NOR-Gate Circuit

Figure 17a shows a second loop circuit, similar to the previous one but with one element (A) operated in a reflection-output mode. This acts as a NOR-gate so that when taken with two OR-gates a loop circuit with overall inversion is created. Because such a circuit automatically inverts the circulating datum on each cycle, no external input is needed to test for proper operation. Figure 17b shows the detected outputs from each gate. The transmission (OR) output was monitored.

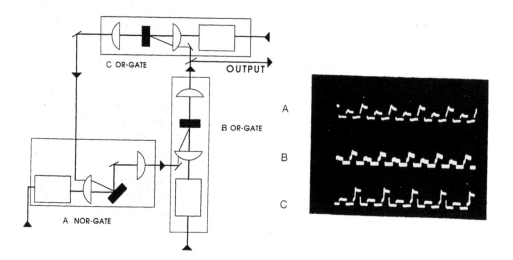

Fig. 17

On reaching gate A, logic-one is rendered logic-zero on one cycle and zero becomes one on the succeeding cycle. Both restoring logic and the inversion function are convincingly demonstrated [18].

Conclusions from Experiments with the Three-Element Loop Circuit

Although the stability of the state-of-the-art NLIFs was limited, the tolerances required and available allowed sufficient output to be derived from the element to drive a succeeding one and with fresh, standard, power supplies ('hold' beams) supplied to each succeeding stage the tolerances also allowed signal restoration. Since the lock and clock data isolation also implies temporary storage, the optical equivalent of a one-channel 'finite state machine' in which data is inputted, processed, stored and recirculated has been demonstrated.

The control was exercised through a microcomputer: this provided a convenient interface to electronics. The optical system and the three-phase clock sequence could apply, as constructed, to a massively parallel logic plane array when these become available. Relatively little change in principle to the optical system is required. A microcomputer controlling such a parallel optical classical finite state machine is in a sense multiplying the capability of the microcomputer by the number of optical channels.

Cascadable All-Optical Signal Gate Full Adder Circuit:
Experimental Realisation

One novel feature of the NLFP is the presence of two almost complementary responses in transmission and reflection. According to a proposal by Wherrett [20], this can be used to give both the SUM and CARRY of a full addition simultaneously using only one active element. The truth table and the form of input/output characteristics is shown in Fig. 18.

If a reflection signal is monitored at four input levels then the response (low-high-low-high) is that demanded of the SUM. Correspondingly the transmission response (low-low-high-high) is that of the CARRY. It is necessary to discriminate between the low levels and the high levels. This is achieved with two further nonlinear plates set at appropriate holding levels. The complete circuit, which has been operated successfully [21], includes a further bistable plate to act as a temporary store. The temporary store also generates gain and reconditions the signal to a sufficiently high power level for the whole process to be iterative. Wherrett suggests schemes whereby this type of gate could be used for optical temporal integration [20].

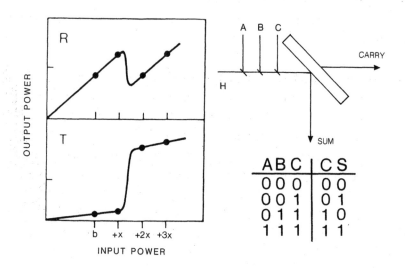

Fig. 18

Optical Circuits with NLIFs at the Optical Sciences Center, Tucson, Arizona

A quite different and complementary set of experiments has been reported over the last several years from OSC. These include the use of a NLIF as a decision maker in an optical circuit for pattern recognition as a way of implementing optical symbolic substitution logic (SSL) [22]. The objective is to identify the number of locations of the pattern formed by two bright spots located next to each other horizontally. This is done by shifting one pixel horizontally and superimposing two shifted patterns with a beam combiner. A single spot in the combined pattern can have one of three light levels: no light, one unit of light or two units of light. The logic decision is made by focussing the combined pattern on a 2 x 3 AND gate array using a fly's eye fibre optic lens array. The output is a bright spot in the right hand pixel wherever the input has bright spots adjacent horizontally. All the locations of this pattern are identified by a single logic operation independent of array size. This experiment has been extended to make a very simple but complete symbolic substitution demonstration. In these experiments one NLIF is used as a decision maker after the pixel shifting optics; in the second part this is cascaded through a second NLIF to perform symbolic scription required in SSL.

In a further experiment the pattern recognition demonstration is extended to two orthogonal shifts to recognise the above spot pattern. This time instead of using a shift input AND gate they cascade two, two input AND gates. In this last paper the OSC group extend the use of an NLIF to an analogue image process application towards associative memory. The NLIF is used as an intensity dependent shutter to control the beam which leads a DCG hologram [23].

The OSC group have also presented the demonstration of all-optical compare and exchange using NLIF devices [24]. This utilises polarisation multiplexing and filtering, and latching, bidirectional logic. The combination of 2-D arrays of compare-and-exchange modules with optical perfect-shuffle connections leads to pipelined optical sorting networks that can in principle process large numbers of high-bandwidth signals in parallel. Optical sorting networks with these characteristics are applicable in telecommunication switches, parallel processor interconnections and database machines.

The Arizona work is characterised by the use of ZnS/ZnSe NLIFs in two-dimensional format (except the last experiment). The experiments are essentially pipelined and have not yet been looped and integrated. In this way they are very complementary to the Heriot-Watt University experiments. The performance of the NLIFs in the Arizona experiments has produced some pessimism from the experimenters. This is partly due to the rather poor performance of these particular early NLIFs. As described earlier, this situation is now improving.

Nonlinear Fabry-Perot Etalons as Data Transparent Routing Switchings

Telecommunications require high-bandwidth rerouting particularly now that the data capacity of optical fibres exceeds that of electronic cross-bar switching technology. All-optical rerouting is also needed in the more modest context of embryonic optical computing architectures. In the optical fibre communications network of the future there will be a demand for spatial switches compatible with the very wide transmission bandwidths of the fibres themselves.

A bistable spatial optical switch has been used to route video data between fibres [25]. In this case a single switch was used in the layout shown in Fig. 19.

The holding beam, control pulses and data (video signal) were all supplied through the input channel - an optical fibre imaged onto the ZnSe NLIF by a graded-index rod lens. Similar lenses were used to couple the transmitted and reflected components into the two output fibres. The 1 x 2 switch was thus powered, set and re-set entirely optically, via the input fibre.

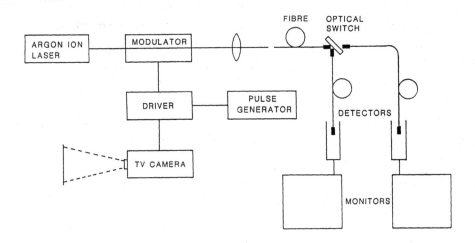

Fig. 19

In this first test experiment, a switch contrast (transmission) of 8.5 dB was demonstrated - sufficient to clearly re-route the video information between display monitors.

In utilising the optical control of bistable interferometers, it was possible to investigate larger networks [26]. A conceptual 1 x 8 network is illustrated in Fig. 20.

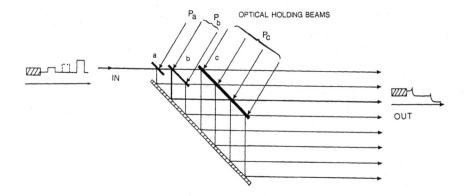

Fig. 20

In this example three bistable plates (a, b and c) are illuminated with focussed beams such that a total of seven independent switches are created; each biased into the bistable regime. Eight possible output paths are available, depending on the transmission state of each node. The bandwidth of these paths is limited only by the width of the interference transmission peak of the F-P device: e.g. 0.5 THz for a 1 nm peak (FWHM) around 830 nm. A specific route through the network is determined by header pulses, preceding the data. These control signals are distinguished from the data signals by their greater energy. If both types of signal are of similar peak power then this corresponds to the control pulses being of longer duration, and to minimise both power and energy requirements, they should ideally last about one switch time. Thus the speed of response of the bistable Fabry-Perot device, determined by the nonlinear medium, is fast enough for it to react to control pulses in, say, the ns - μs range but slow enough to leave it unaffected by data, of similar power level, transmitted at GHz-Thz rates. (The exact dividing line would depend on the nonlinear material and mechanism being exploited). Data pulses of higher peak power could be transmitted through a network provided that the mean power, averaged over one response time does not exceed the switch energy of the most sensitive node.

The factor determining which nodes are switched is the amount of energy in each control pulse. By pre-adjusting the bias level of the bistable switches, relative to their switch powers, each layer of the network can be set to respond only to control pulses above a critical energy: decreasing in magnitude (e.g. by 10-50% steps, depending on tolerances) for each subsequent switch layer. The header is then encoded such that the more energetic pulses (i.e. pulses with higher peak power or longer duration) are used to switch the earlier nodes of the network. The finite switching time of each node is exploited to ensure that the switch-pulse energy, corresponding to one particular node, that is passed on to subsequent (more sensitive) nodes, is insufficient to induce further switching. Once a switch is activated, its bistable response ensures that it latches into its new state so that the subsequent data stream can follow the path set up through the network. Finally, before setting up a new routing, the network would be re-set either by temporarily interrupting the holding beam power to all the nodes or, if they are biased at a level only a little above the switch-off power, by reducing the average (carrier) light level passing through the network.

To demonstrate the concepts underlying routing through a multi-node network, in which coded header-pulses have to be interpreted, the 2-layer (1 x 4) network (corresponding to plates a and b in Fig. 20) was constructed. This was again conveniently set up using available ZnSe NLIF devices operating at 514 nm wavelength (argon-ion laser)(27). It should be noted that other optothermal Fabry-Perot devices, working at 830 nm [17, 28] or

1.3/1.55 μm, and similar bistable Fabry-Perot etalons - based on optoelectronic nonlinearities in GaAlAs (830 nm) [9, 11] or InGaAsP (1.3/1.55 μm) [29], could equally be exploited in this way.

Figure 21 shows the 1 x 4 spatial switch assembly - fabricated from just two coated glass substrates, cemented together. The overall experimental arrangement is shown in Fig. 22.

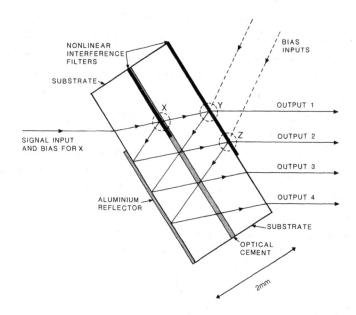

Fig. 21

The output from the laser was split into bias and signal beams: the latter being modulated by an acousto-optic modulator with header pulses of 5 ms duration (comparable to the response time of the devices) plus higher frequency data. The two separate bias beams for nodes Y and Z were produced from a dichromated-gelatin holographic optical element, while node X was biased through the signal input channel. The beam spot-size on each bistable element was 70 μm ($1/e^2$ diameter), giving switch powers of 120 mW. (Much lower switch powers, e.g. a few milliwatts, can be obtained using smaller spot sizes and better optimised, or alternative, nonlinear Fabry-Perot devices.) Nodes Y and Z were biased to within 30 mW of switch power and node X within 15 mW. The switch network was tested by monitoring all the output channels simultaneously, whilst applying the four different header codes to the input channel - followed in each case by a burst of data. The results are shown in Figure 23.

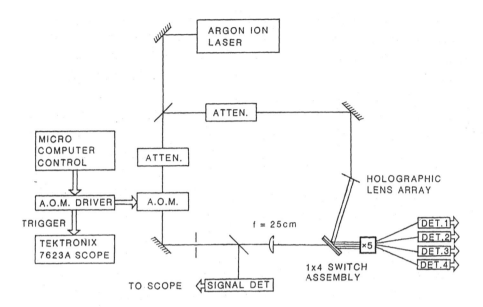

Fig. 22

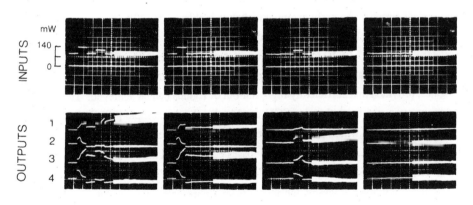

5 ms/div.

Fig. 23

Although in this test the individual switch contrasts are not optimum, the basic concept was demonstrated. That is, without any electrical or mechanical intervention, (i) the required combination of node states could be set by the corresponding header code; (ii) this configuration was seen to correctly latch; and (iii) the subsequent transmitted data had the highest amplitude on the designated output channel.

The network transmissivity to the output channels was not identical for all routes, due to differences between the reflectivities and transmissivities of each node. To allow for this, attenuators were used to balance each channel to give approximately equal output data amplitudes. The overall worst- case loss in this experimental set-up (channel 1) was 16 dB, again as a result of using available (unoptimised) NLIF devices. The highest throughput channel (4) suffered a loss of 3 dB, neglecting the additional attenuation.

Optical Wiring: Interconnects

The circuit concepts discussed in these lectures have implied multiple free space optical interconnections as well as an array of power beams. Conditioning of optical beams to provide, for example, uniform intensity of an array is also a necessity. Since we have indicated that low switching energies dictate micron sized logic elements, it also follows that we concerned with geometrical micro-optics.

Holographic optical elements (HOEs) in various forms can provide such optical wiring for digital optical circuits. An early example is that of Jenkins et al [30] who have used a 4 x 4 array of subholograms to interconnect 16 optical gates (formed on a liquid-crystal-light-valve, LCLV) to create an oscillator plus master-slave flip-flop optical digital circuit. Such a space-variant hologram can provide any arbitrary interconnection between gates, provided that sufficiently high, fan-out of signals can be achieved. The 5% efficient computer generated hologram CGH used in the above example [30] limited the maximum fan-out to 3, at which point the gate inputs corresponded to only 1.7% of their output power levels, even assuming no optical losses elsewhere in the system.

Such an efficiency is extremely limiting. In work at Heriot-Watt University [31], use has been made of volume holograms fabricated from dichromated gelatin (DCG). Such elements can readily have larger than 95% efficiency - this can also be the case when the less efficient alkali halide CGHs are copied onto DCG. The individual lenslets are made by a computer controlled step-and-repeat process. Such an array to provide power beams has already been demonstrated for a 56 x 56 (3136) element array. This sub-system requires a further optical element to provide a transformation from the Gaussian input beams into a flat top intensity profile: again this can be done with conventional lens systems and copied onto DCG HOEs or approached from a CGH.

Multiple exposure allows efficient fan-out and fan-in holograms to be made: a fan-out by a factor of 5 has already been used by us in the demonstration of digital edge extraction [32].

Such HOEs are only some examples of a new generation of micro-optical components: other materials such as silicon nitride can be used to make phase relief Dammann gratings which use diffraction properties to give exactly equal arrays of beams. Combinations giving adequate uniformity and efficiency must be sought. In general the prospects for this new generation of optical elements seem to be excellent.

Conclusion: Future Optical Computing Demonstration Systems

The optical circuit work that has been reviewed in these lectures has shown that cascadable restoring digital optical logic is practicable and, if combined with sufficient parallelism with appropriate architectures, could be significant in computational technology. A useful next step will be to construct a demonstrator optical computing system which recognises the limited parallelism that will be available on a short time-scale but uses components and sub-system architectures that can be shown to be scaleable to approach significant computational problems. Power and complexity questions suggest that a first stage should include about 10 cascadable steps and a limited degree (say 5:1) of fan-out. A proposal from us is called an optial cellular logic image processor (O-CLIP). This is a combination of iterative lock and clock circuits with parallel interconnected pipelines similar to those used in the edge extractor. It is based on a 2-D array of processing cells each receiving two inputs and producing two outputs. A block diagram of a possible configuration of the system is given in Fig. 24.

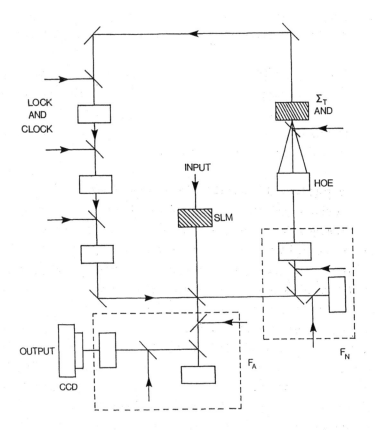

LOCK
AND
CLOCK

Σ_T
AND

INPUT

HOE

SLM

OUTPUT

CCD

F_A

F_N

Fig. 24

It includes a programmable logic function generator involving the interaction of two logic planes - this function has already been confirmed experimentally for a single channel. A selection of computational problems could be addressed by such a demonstrator which adopts a strategy of parallel architecture targeted towards image processing. The term image is to be treated in a general sense as it may well be a 2-D representation of a physical problem instead of an incident spatial image. The attempted construction and operation of such a device would provide a logical next step in developing the technologies described towards significant progress. The proposition is essentially generic in that any nonlinear mechanism for the logic plane devices could be incorporated as future progress indicates.

In assessing the present state of progress the reader is invited to note that, as yet, very little technological development effort has reached the subject: comparison with very early laboratory stages of the triode or transistor is appropriate.

The author acknowledges the use of previously unpublished material which has been provided by colleagues in the Physics Department.

REFERENCES

1.   A.A. Sawchuk and T.C. Strand, Proc. IEEE, 72, 1984.

2.   C.A. Mead and L.A. Conway, "Introduction of VLSI Systems", Addison-Wesley (1980).

3.   M.E. Prise, N. Streibl and M.M. Downs, Opt. and Quan. Electron. 20, 49, 1988.

4.   B.S. Wherrett, IEEE J. Quantum Electron. QE-22, 646, 1984.

5.   N. Bloembergen, "Nonlinear Optics", W.A. Benjamin, (1965).

6.   D.A.B. Miller, M.H. Mozolowski, A. Miller and S.D. Smith, Opt. Commun. 27, 133, 1978.

7.   S.D. Smith & F.A.P. Tooley, in "Optical Bistability 2", ed. C.M. Bowden, H.M. Gibbs and S.L. McCall, Plenum, 1984.

8.   D.A.B. Miller, C.T. Seaton, M.E. Prise and S.D. Smith, Phys. Rev. Lett. 47, 197, 1981.

9.   Y.H. Lee, A. Chavez-Pirson, S.W. Koch, H.M. Gibbs, S.H. Park, J. Morhange, A. Jeffrey, N. Peyghambarian, L. Banyai, A.C. Gossard and W. Wiegmann, Phys. Rev. Lett. 57, 2446, 1986.

10. J.L. Jewell, A. Scherer, S.L. McCall, A.C. Gossard and J.H. English, Appl. Phys. Lett. 51, 94, 1987.

11. R. Kuszelewicz, J.L. Oudar, R. Azoulay, J.C. Michel and J. Brandon in "Optical Bistability IV", eds. W. Firth, N. Peyghambarian and A. Tallet, Editions de Physique, Les Ulis, France, 1988.

12. A.C. Walker, F.A.P. Tooley, M.E. Prise, J.G.H. Mathew, A.K. Kar, M.R. Taghizadeh, and S.D. Smith, Philos. Trans. R. Soc. London A313, 249, 1984.

13. F.V. Karpushko and G.V. Sinitsyn, J. Appl. Spectros, (USSR) 29, 1978.

14. G.R. Olbright, N. Peyghambarian, H.M. Gibbs, H.A. Macleod and F. van Milligen, Appl. Phys. Lett. 45, 1984.

15. E. Abraham, C. Godsalve and B.S. Wherrett, J. Appl. Phys. 64, 21, 1988.

16. A.D. Lloyd and B.S. Wherrett, Appl. Phys. Lett. 53 460, 1988.

17. A.C. Walker, Opt. Commun., 59, 145, 1986.

18. S.D. Smith, A.C. Walker, F.A.P. Tooley and B.S. Wherrett, Nature 325 6099, 1987.

19. N.C. Craft, Appl. Opt. 27 1764, 1988.

20. B.S. Wherrett, Opt. Commun. 56 87, 1985.

21.   F.A.P. Tooley, N.C. Craft, S.D. Smith and B.S. Wherrett, Opt. Commun., 63 365, 1987

22.   M.T. Tsao, L. Wang, R. Jin, R.W. Sprague, G. Giglioli, H.M. Kulcke, Y.D. Li, H.M. Chou, H.M. Gibbs, N. Peyghambarian, Opt. Engin. 26 41, 1987.

23.   L. Wang, V. Esch, R. Feinleib, L. Zhang, R. Jin, H.M. Chou, R.N. Sprague, H.A.Macleod, G. Khitrova, H.M. Gibbs, K. Wagner and D. Psaltis, Appl. Opt. 27 1715 1988.

24.   R. Jin, C.W. Stirk, G. Khitrova, R.A. Athale, H.M. Gibbs, H.M. Chou, R.W. Sprague and H.A. Macleod, IEEE J. Select. areas in Commun. 6 1273, 1988.

25.   C.R. Paton, S.D. Smith and A.C. Walker, in "Photonic Switching" Series in Electronics and Photonics 25, Springer Verlag 1987.

26.   G.S. Buller, C.R. Paton, S.D. Smith and A.C. Walker, Appl. Phys. Lett. 53 2465 (1988).

27.   S.D. Smith, J.G.H. Mathew, M.R. Taghizadeh, A.C. Walker, B.S. Wherrett and A. Hendry, Opt. Commun. 51, 357 (1984).

28.   G.S. Buller, C.R. Paton, S.D. Smith and A.C. Walker, Optical Bistability IV (Optical Society of America, Washington, D.C.) p56 (1988).

29.   H. Kawaguchi and Y. Kawamura, Electronics Letts. 23, 1013 (1987).

30.   B.K. Jenkins, A.A. Sawchuk, T.C. Strand, R. Forchheimer and B.H. Soffer, Appl. Opt. 23 (19) 3455 - 3464 (1984).

31.   A.C. Walker, M.R. Taghizadeh, J.G.H. Mathew, I. Redmond, R.J. Campbell, S.D. Smith, J. Dempsey and G. Lebreton, Opt. Engin. 27, 38, 1988.

32.   F.A.P. Tooley, B.S. Wherrett, N.C. Craft, M.R. Taghizadeh, J.F. Snowdon and S.D. Smith, C2-459 J. de Physique 49 459 (1988).

PHOTOREFRACTIVE DEVICES AND APPLICATIONS

H. Rajbenbach and J.P. Huignard

Laboratoire Central de Recherches, Thomson-CSF

Orsay, France

1.

INTRODUCTION

The development of real time devices for the processing of two-dimensional time-varying coherent wavefronts is a major interest for those working in the field of optical computing. It was recognized early that the photoinduced index change in electro-optic crystals (photorefractive effect) exhibits unique capabilities that allow highly nonlinear functions to be implemented with the use of low-power cw lasers. A wide variety of prototype operations has already been performed, including holographic storage, image amplification and correlation, binary two-dimensional logic, spatial light modulation, associative recall, novelty filtering and beam steering. Besides, numerous media display photorefractive properties at U.V., visible, or infrared wavelengths. Current photorefractive materials consist of electro-optic oxides such as $LiNbO_3$, $BaTiO_3$, $Bi_{12}SiO_{20}$ (or BSO), $Bi_{12}GeO_{20}$ (or BGO), $Ba_{1-x}Sr_xNb_2O_6$ (or SBN), and semiconductors such as GaAs, InP and CdTe. This paper provides a review of theoretical and experimental work on the photorefractive effect in view of pratical applications in the field of optical computing. The following section presents the mechanism of the photorefractive grating recording and the relevant parameters for optical processing applications. In section 3, the conditions for efficient wave mixing in photorefractive media are analyzed. Various applications, framed in the context of optical computing are described in the last sections.

2.
PHOTOREFRACTIVE EFFECT AND MATERIALS

2.1. Introduction[1-3]

The light-induced changes of refractive indices in electro-optic crystals are based on the spatial modulation of charges by nonuniform illumination (fig.1). Electrons (or holes) are photoexcited from impurity centers present in the material and upon migration, are retrapped at other locations, leaving behind positive or negative charges of ionized trap centers. The photoexcited charges will be reexcited and retrapped until they finally drift out of the illumination region. The resulting space-charge field between the ionized donor centers and the trapped charges modulates the refractive indices via the electro-optic effect. Uniform illumination erases the space-charge fields and brings the crystal back to its original state (optical erasure).

The complete mathematical description of grating formation in photorefractive crystals has been derived by Kukhtarev et al.[4] From this model, the photoinduced index modulation at saturation regime is given by the following expression :

$$\Delta n = 2 \, n_0^3 \, r \, \frac{\beta^{1/2}}{1 + \beta} \left[\frac{E_0^2 + E_D^2}{(1+E_D/E_q)^2 + (E_0^2/E_q^2)} \right]^{1/2} \qquad (1)$$

and the spatial phase shift ψ between the incident fringe pattern and the photoinduced index modulation is given by :

$$tg \, \psi = \frac{E_D}{E_0} \left[1 + \frac{E_D}{E_q} + \frac{E_0^2}{E_D E_q} \right] , \qquad (2)$$

where r is the electro-optic coefficient, n_0 is the background index of the medium, β is the incident intensity ratio of the two interfering beams, E_0 is an externally applied field, E_D is the diffusion field and E_q

is the maximum field which would correspond to a complete separation of the positive and negative charges by one grating period. The expressions for these fields are the following:

$$E_D = 2\pi kT/e\Lambda ; \qquad E_q = eN_A \Lambda/2\pi\varepsilon_0 \varepsilon,$$

where N_A is the trap density in the crystal volume, T is the temperature, e is the electron charge, ε_0 is the free space permitivity, Λ is the fringe spacing and ε is the relative static dielectric constant.

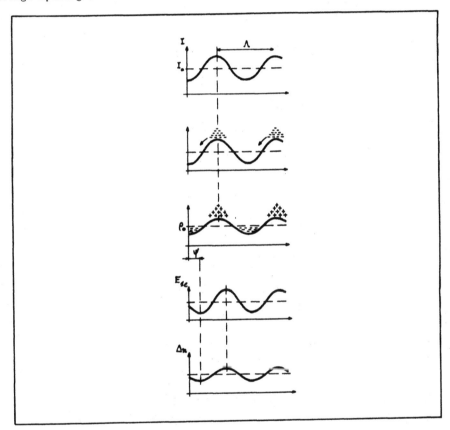

Fig.1: Mechanism of the photorefractive grating recording in electro-optic crystals: light with a spatially periodic intensity resulting from the interference of two beams in the material (I) rearranges the charge density (ρ) that causes a periodic field (E_{sc}) by Poisson's equation. This electric field then causes a change in the refractive index of the crystal (Δn) by the linear electro-optic effect.

3.2. Characteristic parameters of photorefractive crystals

In this section we will briefly discuss some of the most important parameters involved in applications of photorefractive crystals to optical processing i.e, the crystal sensitivity,[5-6] the grating build-up time constants and the steady-state diffraction efficiency.

a) Crystal sensitivity

The photorefractive sensitivity is defined as the refractive index change Δn per unit absorbed energy density :

$$S = \Delta n / \alpha I_0 \tau,$$

where α is the crystal absorption coefficient and recording wavelength λ, τ is the crystal response time and I_0 the incident power density. This definition is a useful figure of merit to compare materials having different absorption coefficients at a given wavelength.[7] The response time of a photorefractive crystal is the dielectric relaxation time multiplied by a function of differents parameters such as applied field E_0, grating spacing Λ, drift and diffusion lengths (respectively r_E and r_D) of the photocarriers :

$$S = \frac{1}{2} n_0^3 \frac{r}{\varepsilon \varepsilon_0} F (E_0, \Lambda, r_E, r_D). \qquad (3)$$

Since n_0 and r/ε are nearly constant for all electro-optic inorganic crystals, the photorefractive sensitivity is mainly determined by the recording conditions and by the relative values of the drift and diffusion lengths compared to the grating spacing. S reaches a maximum value when the excited photocarriers drift of diffuse over distances equal to or larger than the grating spacing. The upper limit of S for an elementary grating ($\beta = 1$) and unit quantum efficiency is:[7]

$$S_{max} = n_0^3 re\Lambda / 4\pi \varepsilon \varepsilon_0 h\nu$$

This can be estimated to 0.1 $cm^3.J^{-1}$ for $\lambda = 0.5$ μm. This optimum photorefractive sensitivity is reached in efficient photoconductive crystals such as KTN, BSO, BGO and GaAs.

Another figure of merit which is commonly used for experiments with photorefractive crystals is the energy per unit area W to write an elementary grating ($\beta = 1$) having 1% or a few percent efficiency in a crystal having 1 mm (or a few millimeters) thickness. This figure of merit enables a ready comparison of slow and fast materials illuminated with the same incident beam intensity I_o.[7] Writing energy as low as $W \sim 100$ µJ.cm^{-2} are available in photorefractive crystals. It must be noted that these values of recording energy in dynamic materials are nearly equivalent to high resolution silver halide plates.

b) Steady state diffraction efficiency

The diffraction efficiency η of a thick phase transmission grating with a peak to peak index modulation $2\Delta n$ is derived from the Kogelnik formula:[8]

$$\eta = \exp(- \alpha d/\cos\Theta).\sin^2(\pi d\Delta n/\lambda\cos\Theta)$$

where Θ is the Bragg angle inside the crystal and d is the thickness of the crystal. High values of the photoinduced index change and therefore of the diffraction efficiency are obtained in materials with high electro-optic coefficients such as ferroelectric crystals BaTiO$_3$,[9] SBN[10] ($r \simeq 10^3$ pm/V) or KNbO$_3$[11] ($r \simeq 60$ pm/V). In other materials having low electro-optic coefficients such as BSO,[12] BGO, GaAs or InP[13-14] ($r \simeq 1$-3 pm/V), Δn can be increased with an externally applied electric field E_0 until saturation occurs for $E_0 = E_q$. As an example, Fig. 2 shows the diffraction efficiency of a crystal of BGO as a function of the applied field.[15] Almost 100 % efficiency is measured for an optimized fringe spacing of 20 µm.

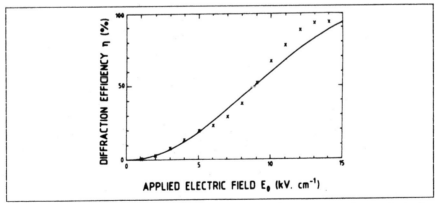

Fig.2: High diffraction efficiencies are obtained with photorefractive volume gratings. In BGO, almost 100 % is measured when a high external dc electric field is applied on the crystal (drift mode). Grating spacing $\Lambda = 20$ µm; beam ratio $\beta = 1$; crystal thickness $d = 10$ mm; wavelength $\lambda = 0.514$ µm. From (15).

c) Response time of the photorefractive effect

The time constant for buidup of a grating is also an important characteristic of the photorefractive effect. The refractive index changes are due to electro-optic effects driven by space-charge fields and the time required to record a grating depends on the efficiency of the charge generation and transport process. The inertia in the nonlinear response of photorefractive media constitutes an important difference from other nonlinear media where the refractive index change is of electronic origin and thus occurs instantaneously. Under continuous wave illumination the crystal response time is given by:[7]

$$\tau = \tau_{di} \frac{(1 + \tau_R/\tau_D)^2 + (\tau_R/\tau_E)^2}{[1+(\tau_R\tau_{di}/\tau_D\tau_I)](1+\tau_R/\tau_D)+(\tau_R/\tau_E)^2(\tau_{di}/\tau_I)} \quad , \qquad (4)$$

where τ_{di} is the dielectric relaxation of the crystal:

$\tau_{di} = \varepsilon\varepsilon_o/n_0\mu e$;

n_0 is the free carrier concentration due to the incident illumination I_0:

$n_0 = \tau_R \alpha \phi I_0/h\nu$,

μ is the mobility of the photocarriers and ϕ the quantum efficiency. The charge recombination time τ_R may be written as:

$\tau_R = (\gamma_R N_A)^{-1}$,

where γ_R is the recombination coefficient, τ_E and τ_D are the drift and diffusion time of the charges, given respectively by:

$\tau_E = 1/K\mu E_0$, $\tau_D = 1/\mu k T K^2$.

$K = 2\pi/\Lambda$ and τ_I is the inverse of the sum of photogeneration rate sI_0 and ion recombination rate $\gamma_R n_0$:

$\tau_I = (sI_0 + \gamma_R n_0)^{-1}$.

A simple expression for the time dependence of the space-charge field during grating recording is the following:

$\Delta E_{sc} = m E_{sc} [1 - e^{-t/\tau}]$.

During erasure by uniform illumination, the photoinduced space-charge field decreases according to the relation:

$\Delta E_{sc} = m E_{sc} e^{-t/\tau}$,

where E_{sc} is the initial amplitude of the field and m the incident modulation. The recording erasure cycle is therefore symmetrical as shown in fig.3 for photorefractive BSO crystals. The typical recording erasure time for an elementary grating of 10-100 ms corresponds to an incident intensity of 10-100 mW.cm^{-2} at the blue or green line of the Argon laser. BSO and GaAs are used as fast and sensitive materials, while crystals such as BaTiO$_3$ have a large electro-optic coefficients but respond rather slowly, i.e, response times of few seconds. Therefore, another important figure of merit of a photorefractive crystal will be the energy required to reach the steady-state diffraction efficiency and this parameter often determines the crystal chosen for a particular application.

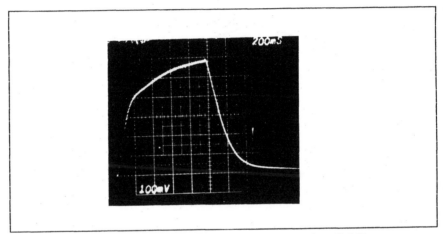

Fig.3: Grating recording-erasing cycle in BSO at λ = 0.568 μm. Incident beam intensity 120 mW.cm^{-2}; E$_0$ = 6 kV.cm^{-1}; Fringe spacing Λ = 6 μm. Steady state diffraction efficiency η = 10% monitored with a low power He- Ne laser. From 16.

3.

BEAM COUPLING IN PHOTOREFRACTIVE CRYSTALS

3.1. Introduction

The recording of phase volume gratings in photorefractive media leads to a stationary energy exchange between the two interfering beams. The resulting energy redistribution that has been observed in many electro-optic crystals (LiNbO$_3$, KNbO$_3$, BaTiO$_3$, BSO and GaAs)[3,16,17] is due to a self-diffraction process of the reference pump beam by the dynamic phase grating photoinduced in the crystal. More specifically, the self-interference of the incident beam with the diffracted beam creates a new holographic grating which can add to (or subtract from) the initial one. Since the diffracted wave is phase delayed by π/2 with respect to the reading beam, the maximum energy transfer is obtained when the incident fringe pattern and the photoinduced index modulation are shifted by $\psi = \pi/2$.[1-3,18]

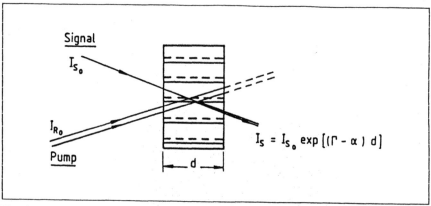

<u>Fig.4</u>: Degenerate two-wave mixing (2WM) in photorefractive crystals. The continuous and dashed lines represent the maxima of the illumination and index pattern, respectively. In the diffusion mode (no applied field), the phase shift ψ is $\pi/2$ and large gains are observed in crystals such as $BaTiO_3$ or SBN.

In photorefractive crystals, such a $\pi/2$ phase shift exists when the recording is by "diffusion" of photocarriers (no external applied electric field) as shown in Fig.4. As a consequence, a permanent and efficient amplification of a low intensity signal beam has been observed in crystals like $BaTiO_3$, $LiNbO_3$, or $KNbO_3$. If we now apply the coupled wave equations to the $\pi/2$ phase shifted component of the photoinduced index modulation, the coherent interaction between the two waves of respective amplitude R and S is described by the following equations:[8]

$$\frac{dS}{dz} = \frac{1}{2}\,\Gamma\,\frac{R^2 S}{R^2+S^2} - \frac{1}{2}\,\alpha R \;\; ; \quad \frac{dR}{dz} = \frac{1}{2}\,\Gamma\,\frac{RS^2}{R^2+S^2} - \frac{1}{2}\,\alpha S \;\; , \tag{5}$$

and the transmitted signal beam intensity resulting from the dynamic two-beam coupling takes the form :

$$I_s = I_{s0}\,\frac{\beta + 1}{\beta + \exp(\Gamma l)}\,\exp\left[(\Gamma - \alpha)d\right] \;\; , \tag{6}$$

where Γ is the exponential gain coefficient of the interaction. Γ is related to the maximum amplitude of the photoinduced index modulation Δn_s through the relation :

$$\Gamma = \frac{4 \pi \Delta n_s}{\lambda \cos \theta} \sin \psi \ . \tag{7}$$

ψ is the spatial phase shift of the grating and, in agreement with the previous arguments, Γ is maximum when $\psi = \pi/2$. In the case of a negligible pump beam depletion, the transmitted signal beam intensity is simply given by :

$$I_s = I_{s0} \exp \left[(\Gamma - \alpha)d \right] , \tag{8}$$

and therefore, when the condition $\Gamma > \alpha$ is fulfilled, the incident signal exhibits gain and the photorefractive crystal may be regarded as a parametric amplifier.

A practical parameter for characterizing the energy transfer due to the two-beam coupling is the effective gain γ_0 defined by the ratio:[19]

$$\gamma_0 = \frac{I_s \ (\text{with pump beam})}{I_s \ (\text{without pump beam})} , \tag{9}$$

and for the undepleted pump beam approximation we have :

$\gamma_0 = \exp (\Gamma d)$.

Therefore, from the measurement of γ_0, the value of the exponential gain coefficient Γ of the interaction can be easily deduced. Large values of Γ ($\Gamma \simeq 20$ cm^{-1}) may be obtained in materials having large Δn_s when recording by diffusion ($\psi = \pi/2$) and this is the case for $BaTiO_3$, $LiNbO_3$ and $KNbO_3$. However, the same 2WM experiment performed with highly photoconductive BSO (or BGO) crystals leads to very low beam coupling ($\Gamma \simeq 1.5$ cm^{-1}) for the following reasons: (i) For diffusion, the required phase shift $\psi = \pi/2$ is established but the steady-state index modulation is low and (ii) For drift, the index modulation is much higher, but the corresponding phase shift ψ is negligible. However, efficient beam coupling can be obtained if the fringe pattern (or the crystal) is moved at a constant velocity with an electric field applied to the crystal. The speed is adjusted such that the index modulation is recorded at all times, but with a spatial phase shift with respect to the interference fringes. Clearly, the optimum fringe velocity will depend on the recording erasure time constant τ and when an interference pattern moving at velocity v is introduced into the coupled wave equations, the resulting gain coefficient Γ is given by:[20]

$$\Gamma = \frac{4 \pi \Delta n_s}{\lambda \cos \theta} \frac{K v \tau}{1 + K^2 v^2 \tau^2} .$$

In a 2WM configuration such as that shown in Fig.5, the fringe displacement increases the amplitude of the $\pi/2$ phase shifted component of the index modulation and consequently efficient energy transfer is obtained in photorefractive crystals like BSO[16] and GaAs.[17] The expected optimum fringe velocity is $v_0 = \Lambda (2 \pi \tau)^{-1}$ which corresponds to a frequency detuning by $\delta \omega = \tau^{-1}$ of the reference beam.[21] This interaction with moving fringes is named "nearly-degenerate two wave-mixing". A simple method for frequency detuning the reference beam by $\delta \omega$ is to use a piezo-mirror driven by a saw-tooth voltage.[22] Fig.6a shows the resonance of the gain on the fringe velocity in GaAs.

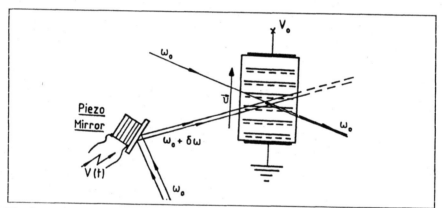

Fig.5: Nearly-degenerate two-wave mixing in BSO or GaAs. The pump beam is frequency shifted by a reflection off a piezo-mirror and the fringes move at velocity v.

3.2. Influence of the recording parameters (spatial frequency, fringe velocity, beam ratio)

A precise knowledge of the spatial frequency response of photorefractive crystals is important for applications to coherent image amplification and optical signal processing.

Fig.6b shows the dependence of the GaAs nearly-degenerate 2WM gain Γ on the fringe spacing Λ and as a function of the applied voltage V_0. For each measurement the fringe velocity is adjusted such that the maximum of gain is obtained. The incident pump beam intensity is 40 mW.cm^{-2} at the recording wavelength $\lambda = 1.06$ µm (diode-pumped YAG laser) and the incident beam ratio is $\beta = 10^3$ (corresponding to a time constant $\tau \simeq 120$ ms). These curves show a strong increase in the gain for $\Lambda \sim 20$ μm. Fig.7 represents the variation of Γ as a function of the incident beam ratio β.

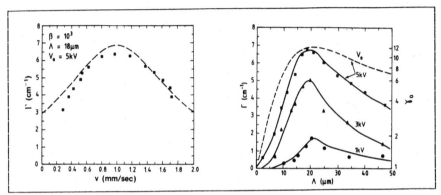

Fig.6: Evolution of the gain coefficient in GaAs at $\lambda = 1.06$ μm.
(left) Large gains occur when the fringes move at the optimum velocity: the space-charge field is optimized and $\pi/2$ shifted with the illumination.
(right) Exponential gain coefficient as a function of the grating fringe spacing for different applied voltages. $\beta = 10^3$. Dashed lines: theoretical plots. From (17).

The following points summarize the main conclusions that can be drawn from these curves : (i) High gain is available in photorefractive semiconductor GaAs when recording with a high electric field and moving the fringes at the optimum velocity. (ii) The gain of the amplifier is strongly dependent on the grating spatial frequency. (iii) The gain reaches saturation at high beam ratio. Consequently, a wide range of experimental conditions allows one to obtain a value of Γ in excess of the crystal absorption losses ($\alpha \simeq 1.6$ cm^{-1} at $\lambda = 1.06$ μm for GaAs) and $\Gamma \simeq 6$-7 cm^{-1} for optimized recording conditions.

The dependence of Γ on the grating spatial frequency Λ^{-1} and the incident beam ratio β are described by the Kukhtarev's equations. Indeed, fitting the experimental plots with the theoretical predictions yields a set of numerical values for crystal parameters such as the mobility and the recombination coefficients. The starting point is the set of differential equations that describe the charge transport and trapping and for which the incident fringe illumination is :

$$I\ (x,t) = I_0\ [1 + m \cos K\ (x-vt)]\ .$$

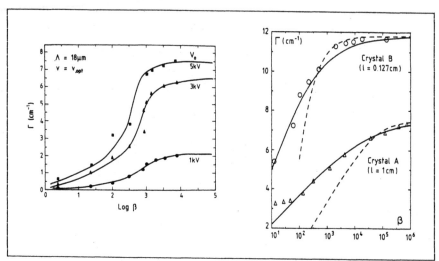

<u>Fig.7</u>: Dependance of the gain coefficient in GaAs at λ = 1.06 μm
(left) and in BSO at λ = 0.568 μm (right) on the incident beam
ratio $\beta = I_{R0}/I_{S0}$. The nonlinear gain is explained by
introducing the second order term in the expansion of the
space-charge field. From(16-17).

The details of the theory may be found in Refs.(23-24), and therefore
we just emphasize in this section the important features related to the
fringe displacement in crystals like GaAs and BSO. Derived from these
references, the velocity that maximizes the imaginary part of the space
charge field responsible for energy transfer is:

$$V_{opt} = s (I_{R0} + I_{S0}) \frac{N_D}{N_A} \frac{E_q}{E_0} \frac{\Lambda}{2\pi} , \qquad (11)$$

where s is the ionization cross-section, N_D the density of donor atoms and
N_A the density of acceptor atoms.

An optimum grating spacing exists and can be found from the condition
$E_M E_q = E^2_0$ where $E_M = \gamma_R N_A \Lambda/2\pi\mu$, leading to :

$$\Lambda_{opt} = \frac{2 \pi E_0}{N_A} \left[\frac{\mu \varepsilon_0 \varepsilon}{\gamma_R e} \right]^{1/2} , \qquad (12)$$

and to an optimum frequency detuning of the pump beam by

$$\delta\omega = 2 \pi v_{opt} \cdot (\Lambda^{-1}_{opt}) = s I_{R0} \frac{N_D}{N_A} \frac{E_q}{E_0} . \qquad (13)$$

It may be concluded that recording in GaAs with a moving grating has two consequences. First under optimum conditions (v_{opt} and Λ_{opt}) the space-charge field is $\pi/2$ out of phase with the interference pattern, i.e, all the space-charge field is useful for promoting the energy transfer from the reference beam to the low intensity probe beam. Secondly, the modulus of the space-charge field is increased from a value of mE_0 in the absence of fringe movement to $m(E_q/2)$ at the velocity v_{opt} (typically a five to tenfold increase for $E_0 \simeq 10$ kV.cm^{-1}).

The dependence of the gain Γ on the incident beam ratio β is interpreted by introducing the second-order terms in the expansion of the space charge field (second order perturbation). The space charge field becomes of the form:[24]

$$E_{sc} = 1/2\ E_{s1} \exp [iK(x-vt)] + 1/2\ E_{s2} \exp [2iK(x-vt)] + c.c.\ .$$

4. Summary of crystal performance

Fig.8 summarizes some of the properties of different photorefractive crystals which are of interest for optical signal processing applications. In addition to the data concerning the time response, recent observations[25-27] have shown that, once the grating is recorded (after time τ), a differential gain of about $\Gamma/2$ is available at very high speed (10^{-6} sec. and less).

	λ (μm)	τ	Γ(cm^{-1})	R
Ferroelectrics				
LiNbO$_3$, KNbO$_3$ BaTiO$_3$, SBN	0.514	seconds	10–20	1–50
Non-ferroelectrics				
BSO, BGO	0.568	10–100 ms	8–15	1–3
GaAs	1.06	10–100 ms	6–7	1–5

Fig.8: Properties of some photorefractive crystals for cw incident intensity $\simeq$ 10–100 mW.cm^{-2}. τ, time response; Γ, exponential gain coefficient; R, reflectivity in 4WM.
(In KNbO$_3$,[1] and BaTiO$_3$,[28] faster speeds are demonstrated with highly reduced crystals and elevated temperature, respectively)

4.

APPLICATION OF BEAM COUPLING TO OPTICAL COMPUTING OPERATIONS

4.1. Image amplification

The large values of the gain coefficient Γ in photorefractive crystals permit the amplification of a low intensity signal beam containing spatial information (data plane).[29] The optical set-up for image amplification of a signal wavefront modulated by a photographic transparency is shown in Fig.9.

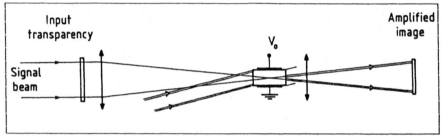

Fig.9: Optical set-up for coherent image amplification. The input signal is introduced in the signal beam path as a spatially intensity modulated wavefront and amplified via 2WM.

With this configuration, the energy transfer from the pump beam allows receipt of an amplified image in the detection plane. When using a photorefractive amplifier such as GaAs or BSO, an electric field is applied on the crystal and the fringe velocity is adjusted in order to receive maximum gain . However, since the spatial frequency response of the photorefractive amplifiers are of the bandpass type, the difference in gain for the various spatial frequencies may be noticeable and can limit the size of the image to be amplified. Fig.10 shows amplified images for $BaTiO_3$ ($\lambda = 0.514\mu m$),[30-31] BSO ($\lambda = 0.568$ μm)[16] and GaAs ($\lambda = 1.06$ μm).[17] Higher values of the gain are possible in BSO/GaAs when the electric field is increased but this would correspond to a loss in image uniformity and quality. Efficient image amplification are performed with photorefractive crystals such as $BaTiO_3$ in which the phase-shifted volume hologram is recorded by diffusion (no applied field) with a carrier spatial frequency of the order $\Lambda^{-1} \sim 1000$ mm^{-1}. To summarize, the main limitations of these coherent image amplifiers stem from crystal inhomogeneities and from light-induced scattering, which limit both the resolution and the minimum intensity of the image to be amplified. This light-induced scattering is due to local fluctuations of the crystal dielectric constant, thus causing scattered waves which are efficiently amplified by 2WM with the pump beam. Clearly, in these experiments, a trade-off between the gain of the photorefractive amplifier and the signal to noise ratio and resolution of the amplified image has to be expected.

Fig.10: Amplified images via 2WM in photorefractive crystals.
        (a) (x1000) in BaTiO$_3$, λ = 0.514 µm (diffusion). From (30).
        (b) (x20)   in BSO, λ = 0.568 µm (drift mode).From (16).
        (c) (x5)   in GaAs, λ = 1.06 µm (drift mode). From (17).

4.2. Amplified phase conjugation in photorefractive crystals

Optical phase conjugation with photorefractive crystals utilizes a four-wave mixing (4WM) interaction to reverse both the direction of propagation and the phase of an arbitrary input wavefront.[32] Phase conjugate mirrors have many applications in problems associated with passing through distording media: the phase distortion can be removed by allowing the wavefront to travel back through this same media (Fig.11).

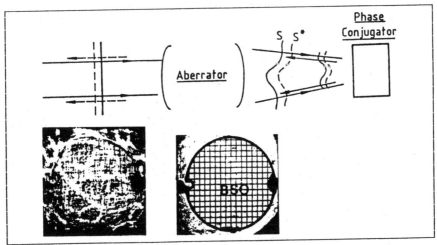

Fig.11: Phase distortion compensation by wavefront reflection on a phase conjugate mirror. The distorted image (bottom left) is restored after travelling back through the same medium. From (33).

A second property of phase conjugate mirrors is their ability to
generate a conjugate signal with an amplified intensity. Amplified phase
conjugation has been observed in photorefractive crystals such as $LiNbO_3$,
$LiTaO_3$, $KNbO_3$, $BaTiO_3$, SBN,[34-37] with typical time responses of a few
seconds, and in BSO[38] and GaAs[39] (10-100 ms) when recording with
moving grating. The optical configuration used for phase conjugation by
nearly-degenerate 4WM is presented in Fig.12. In this interaction, the
conditions of high reflectivity closely depend on the same parameters as
the exponential gain coefficient Γ previously considered in the 2WM
interaction, i.e, the fringes in the crystal move at a constant velocity
and the fringe spacing is adjusted at the optimum value ($\Lambda_{opt} \simeq 20$ μm for
GaAs). Two important extra parameters in this 4WM interactions are (i) the
pump beam ratio $r = I_{P2}/I_{P1}$ and (ii) the polarization of the "reading"
beam I_{P2}. As shown in Fig.13, there is a noticeable dependance of the
conjugate beam reflectivity on these parameters. The lowest curve (R^-) is
obtained for parallel polarisations of the interacting beams. The maximum
of reflectivity ($R^- \sim 1$) obtained for asymmetric pump beam ratio ($r < 1$),
is in accordance with the coupled mode theory developped in Ref.(37).
Higher reflectivities ($R^+ = 5$) are demonstrated with crossed polarization
of the pump beams (upper curve, R^+). This last configuration was first
proposed by Stepanov et al[40] as a mean to take full advantage of the two
orthogonal axes of birefringence available in cubic crystals so as to
provide gain for both the signal beam and the phase conjugate beam.

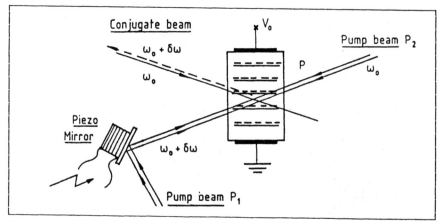

Fig.12: Nearly-degenerate 4WM configuration. The pump beam is
frequency shifted to obtain amplified phase conjugation.

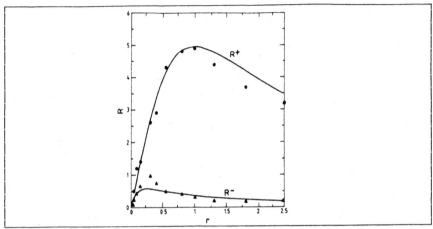

Fig.13: Reflectivity of GaAs phase conjugate mirrors as a function of
the pump beam ratio r = I_{P2}/I_{P1}.
(R⁻) All the beams have the same polarization.
(R⁺) Enhanced reflectivity is obtained with the use of
crossed-polarized pump beams. From(39).

4.3. Laser beam steering / Optical interconnections

This application relies upon the use of a two-dimensional spatial light
modulator in combination with a photorefractive crystal (Fig.14a). The
basic principle is the following : the pump beam interferes in the crystal
with the probe beam which direction is selected by the spatial light
modulator (array of electrooptic shutters for example), and after two-beam
coupling, a complete energy exchange from the pump to the selected probe
beam direction can be obtained by using photorefractive crystals with
large gain coefficients. Therefore we can say that the pump beam has been
deflected in the direction of the probe. If another direction of the probe
beam is selected, the previous grating is erased, and rewriting a new one
deflects the pump beam in another direction. By means to this principle, a
new type of random-access digital laser beam deflector with large scan
angles is realized.

A practical demonstration of this principle over a limited number of
positions is achieved with the experimental set-up shown in Fig.14b. The
low intensity signal beam is expanded and reflected by an array of
piezo-mirrors (4 x 3). In the focal plane of lens L where a
photorefractive $BaTiO_3$ crystal is placed, the pump beam and the array of
signal beams interfere. Selection of one probe beam direction is achieved
as follows: all of the piezo-mirrors are excited with a ramp generator

except for one, which corresponds to the selected direction of deflection.[41] Due to the Doppler shift δ induced by the moving mirrors, the interference fringes move. If $\delta \gg \tau^{-1}$ ($\tau \sim 1$ s. is the time constant for energy exchange in $BaTiO_3$), the corresponding index modulation cannot be recorded due to the crystal inertia. Therefore, the probe beam whose direction is selected by the nonexcited piezo-mirror is amplified. Fig.14c shows the experimental results obtained by driving the piezo-mirrors, where the deflected beam is about 10% of the pump intensity. In this experiment the use of a mirror array with a temporal phase modulation of the incident wavefront makes possible a perfect discrimination between the nonselected and the selected probe beam directions and this deflection principle can certainly be applied for reconfigurable optical interconnection applications.[42-45]

4.4. Self-induced optical cavities

Due to the large gain coefficients and the large reflectivities in 2WM and 4WM, different types of self-starting oscillators can be obtained by adding an optical feedback to the photorefractive amplifiers.[46] These coherent oscillations have been reported with $BaTiO_3$, SBN and $LiNbO_3$ due to the high gain resulting from the high electro-optic coefficients and they are also obtained with BSO[47,48] and GaAs[39] because of the gain enhancement due to self-induced moving gratings when an electric field is applied to the crystal. Some of the characteristic properties of the ring and phase-conjugate oscillators are reviewed in the following sections.

a) Ring oscillators

The optical set-up for obtaining a ring oscillator from a photorefractive amplifier is shown in Fig.15. The photorefractive crystal is introduced into the beam path defined by the three mirrors M_1-M_2-M_3, and the angle between the pump beam and the M_1-M_2 direction is chosen so as to correspond to the optimum fringe spacing for the energy transfer of the pump beam. The condition for oscillation is given by:

$$(1 - R_{BS}) R^3 \exp [(\Gamma - \alpha_t) d] \geq 1 , \qquad (14)$$

where R and R_{BS} are the reflectivities of the cavity mirrors and beam splitter, Γ is the gain coefficient of the 2WM interaction and α_t

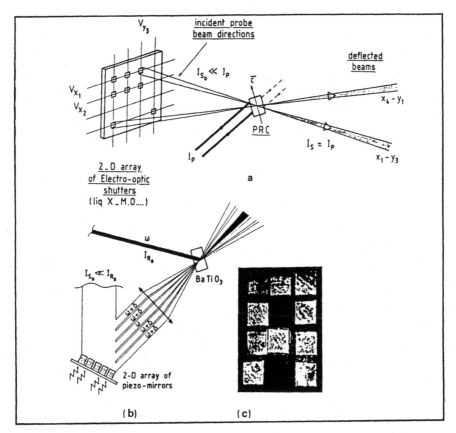

Fig.14: (a)Application of the energy transfer in BaTiO$_3$ to 2D laser
beam steering. (b) laser beam deflection obtained by driving
an array of 4 x 3 piezo-mirrors. (c) Generated pattern. The
beam is ramdomly deflected by driving all the piezo-mirrors
except the ones corresponding to the deflection positions.
From (41).

represents the total losses. Since the values of Γ ($\Gamma > $ 6-7 cm^{-1})
considerably exceed the cavity losses ($\alpha_t \simeq$ 3 cm^{-1}), oscillation builds up
in the cavity.[46] The oscillation in the cavity is self-starting ; the
optical noise due to the pump beam is sufficient to generate a weak probe
beam that is then amplified after each round trip in the cavity. The
required detuning $\delta\omega$ between the pump and the cavity beam in the ring
oscillator is also self-induced. In other words, the crystal chooses from
the optical noise spectrum the frequency component shifted by $\delta\omega$ that will
be optimally amplified in the cavity. In photorefractive GaAs for an
applied voltage Vo = 5 kV, the beam in the cavity is typically frequency
shifted 10-100 Hz for I_{R0} = 40 mW.cm^{-2} at λ = 1.06 μm. A specific property

of these photorefractive ring oscillators is that the gain is unidirectional and only one wave is amplified in the cavity. In particular, the residual coherent retrodiffused beams due to the mirrors M_1, M_2 and M_3 are not amplified : after interference with the pump beam they give reflection type photoinduced gratings that are not efficiently recorded in the GaAs with this configuration. The theory of oscillation in photorefractive ring resonators is developed in Refs.(49,50) and in particular, it is shown that the amount of frequency shift depends on the length of the cavity. Consequently, these photorefractive resonators may be used in a new type of interferometry which directly converts optical pathlength changes into frequency shifts. The peculiarities of these ring cavities can also be applied for the conception of new gyroscopes based on the Sagnac effect.

The specific properties of these oscillators have also been applied to analog optical computing. In particular, if an operator such as matrix-vector multiplier is introduced in the cavity, the feedback loop permits parallel iterative algorithms to be implemented. For example, the inversion of a matrix B can be obtained by calculating the sum $\Sigma (I-B)^n$, each term being provided by a round trip in the photorefractive cavity.[51]

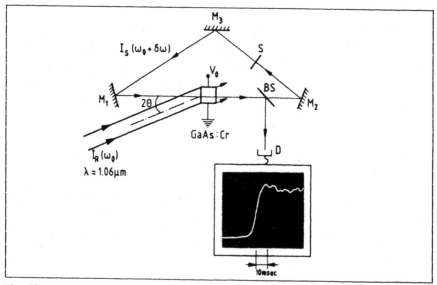

Fig.15: Self-induced optical ring resonator with a photorefractive GaAs.

b) Oscillators with phase conjugate mirrors

As reviewed in Sect.4.2, the conjugate beam reflectivity in a 4WM interaction exceeds unity after optimization of the grating recording parameters. It is thus possible to induce an oscillation between a classical mirror and a photorefractive phase conjugate mirror. Since the first demonstration with a BaTiO$_3$ crystal,[52] similar phase conjugate resonators have been obtained with LiNbO$_3$, and more recently with BSO[46] and GaAs[48] crystals. As shown in Fig.16 for GaAs, the oscillation in the cavity builds up from the noise only when a frequency shift δ is introduced between the pump beams. In such conditions, the beam oscillating in the cavity is frequency shifted by δ/2 and this frequency shift ensures a grating moving at the optimum velocity. The oscillation is maintained even if an aberrator is placed between the mirror M and the phase conjugator crystal.

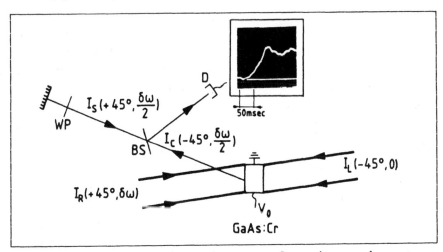

Fig.16: Self-induced oscillation between a plane mirror and a photorefractive GaAs conjugate mirror.

4.5. Optical logic gates and parallel algorithmic state machines

In the area of digital optical computing, two-beam coupling in photorefractive materials offer unique features that suggest the implementation of single instruction multiple data (SIMD) machines. The modules of such a machine consist of large arrays of 1 bit processors executing identical instruction streams in parallel. An important part of a SIMD machine is an optical logic gate array that can be addressed repeatedly, thus yielding practical realization of algorithmic state machines. Optical digital logic such as **OR, AND, NOR** and **NOT** was demonstrated using two beam coupling in photorefractive crystals.[53] In these materials, the nonlinear phenomena rely on the properties of the gain coefficient Γ in a classical two-beam interaction with gain. Three different effects are employed to perform optical logic operations : gain saturation, pump beam depletion, and· optically controlled two-beam coupling. These interactions are detailed in Ref.(53). Example of an OR gate is shown in Fig.17a. The weak signal beam is amplified by the same amount when one or both pump beams I_2, I'_2 (both bearing spatial information) are present with a high intensity level (logic 1). This property, related to the saturation of the photorefractive gain versus the incident pump intensity leads to the logic operation **OR**. The results shown in Fig.17b-c were obtained with photorefractive $BaTiO_3$ crystals and using high/low level transmitance transparencies that spatially modulate the intensities of the pump beams I_2 and I'_2.

In the interaction shown in Fig.18, the signal amplification is controlled by an additional crossed-polarized signal that can erase the interference grating formed by the signal and pump beams. Logic 1 (high intensity level) is obtained only when: (i) Signal and pump beams (I_1 and I_2, respectively) are in logic 1 (high intensity level) and (ii) Control beam I_3 is in logic 0 (low intensity level). Consequently this interaction provides the implementation logic gate $\{(I_1)$ **AND** (I_2) **AND** (**NOT** $I_3)\}$. A parallel **AND** gate $\{(A)$ AND $(B)\}$ can therefore be implemented and combining two crystals, an **exclusive OR** (**XOR**) can also be performed as:

$\{(A)$ **XOR** $(B)\} = \{(A)$ **AND** (**NOT** B)$\}$ **OR** $\{($**NOT** A$)$ **AND** $(B)\}$.

Detailed analysis on the dynamic range of these interactions can be found in Ref.(54). The input images (two-valued intensity bit planes) were introduced with the use of a microcomputer-controlled spatial light modulator. Fig.18 shows the experimental results for input images containing 256 bits in a 16 x 16 format. Moreover, the real-time capability of this device was used for the realization of a photorefractive algorithmic state machine (PASM) that implements a real-time sequential parallel algorithm for binary addition. The technique is to reapeatedly address the parallel (AND, XOR) gates used as a half-adder circuit.[54] The results of the parallel addition of two sequences of 5-bit numbers (i.e., two vectors) is shown in Fig.19.

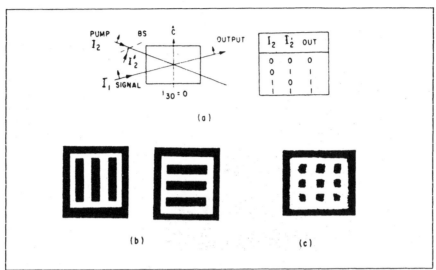

Fig.17: Optical logic OR by two beam coupling in photorefractive crystals.
(a) Principle of operation: high intensity level (logic 1) is obtained when either pump I_2 or I_2' is present.
(b) Input images (A), (B).
(c) Output logic $\{(A)$ OR $(B)\}$. From (53).

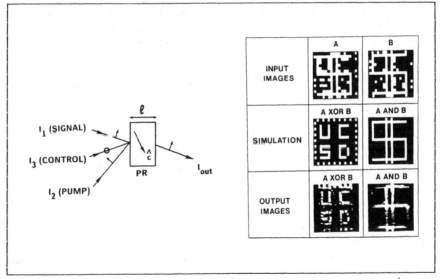

Fig.18: (left): optical logic gate $\{(I_1)$ AND (I_2) AND (NOT $I_3)\}$
I_3 is a control beam that can erases the grating.
(right): a parallel half-adder circuit (XOR and AND) is implemented based on a combination of three $BaTiO_3$ crystals. From (54).

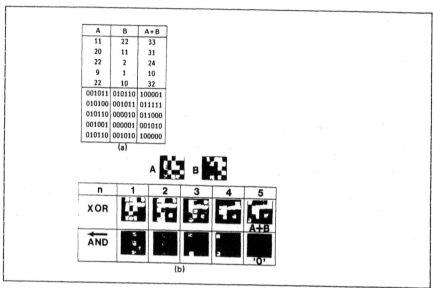

Fig.19: An algorithmic state machine that performs a sequential
algorithm for the parallel binary addition of two vectors A
and B is obtained by repeatedly addressing a photorefractive
parallel half-adder circuit.
(a): Analog and digital versions of the problem.
(b): Experimental results. Top pictures: input bit planes.
lower pictures: outputs of the parallel half-adder circuit for
the five iterations; the carriers propagate and the last
iteration gives the binary representation of the sum A+B.
From (54).

4.6. Image subtraction using a self-pumped phase conjugate mirror
      interferometer

The large electro-optic coefficients of photorefractive crystals such as
$BaTiO_3$ or SBN permit the realization of self-pumped phase conjugate
mirrors.[55,56] While classical 4WM employs two external pump beams (as in
Fig. 13), the pump beams of a self-pumped phase conjugate mirror are
generated via amplified scaterring and interfaces reflections. Self-pumped
phase conjugate mirrors can be used in an interferometric set-up to
perform parallel image subtraction, intensity inversion and exclusive OR
logic operation (Fig.20).[57,58] The incident optical field is divided by
beam splitter BS whose amplitude reflection and transmission coefficients
are r and t, respectively. For the waves propagating in the opposite
directions, the amplitude reflection and transmission coefficients are r'
and t'. Each of the two waves is then passed through a transparency whose
intensity transmittances are T_1 and T_2. These two waves are reflected by a

self-pumped photorefractive phase conjugate mirror with a nearly identical reflectivity R. The phase conjugate beams recombine interferometrically at beam splitter BS to form an output image intensity given by :

$$I_{out} = I_0 \ R \ |t' \ r* \ T_2 + t* \ r' \ T_1|^2.$$

From the stockes principle of reversibility of light it holds that :

$$r't* + t'r* = 0 \ ,$$

and therefore :

$$I_{out} = I_0 R \ |r'r*|^2 \ |T_1 - T_2|^2.$$

Consequently, the interferometer provides an image intensity subtraction proportional to the square of the intensity transmittance functions of the two input slides. This operation represents the Boolean exclusive OR achieved in parallel between the two images T_1 and T_2. The image intensity subtraction occurs in one step. Moreover, the interferometer is only sensitive to intensity difference and is independent of the phase information of the transparencies or optical path lengths of the two arms.

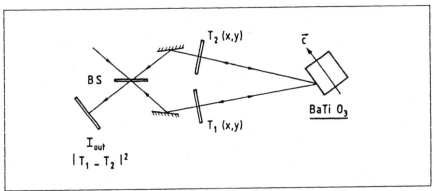

Fig.20: Real time image intensity subtraction using a self-pumped phase conjugte BaTiO_3 crystal. From (57).

4.7. Novelty filters

Self-pumped phase conjugate mirrors can also be employed to perform dynamic filters that detect variations in intensity or phase of a two-dimensional image or data plane.[59] In the experiment shown in Fig.21, two arms of an interferometer share a common phase-conjugate mirror. A transmitting spatial light modulator imposes a phase image onto the optical beam in one of the interferometer arm. The phase conjugate mirror guarantees that, in steady state, the output of the beam splitter is zero, independent of the relative lengths of the arms. However, when the phase information in one arm suddenly changes, the recombined fields at the beam splitter no longer interfere destructively, thus yielding non-zero intensity at the output. The output returns to its original state after a delay governed by the response time of the phase conjugate mirror.

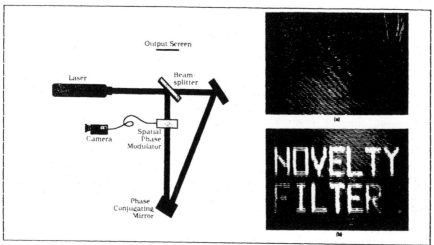

Fig.21: An optical novelty filter based on self-pumped interferometry (left): The spatial light modulator imposes a phase image onto one arm of the interferometer. The output shows time varying features of the impose image, and returns to zero after a time exceeding the time response of the photorefractive crystal. (right): Experimental results : top view : steady state ; bottom view: output taken just after the image is fed in the light valve. From (59).

4.8. Image convolution and correlation

Dynamic cross-correlation or spatial convolution with a classical Fourier transform lens configuration can be achieved by two or four wave mixing of optical fields in photorefractive crystals. A four wave mixing geometry used in these experiments detailled in Ref.(60) is shown in Fig.22. All the beams have the same wavelengths, and the amplitudes $u_1(x,z)$, $u_2(x,z)$ and $u_3(x,z)$ in the outer focal planes are Fourier transformed by propagating to the common focal plane. The transform fields mixed in the photorefractive crystal are the following :

$$u_1 = FT [u_1(x,z)] \; ; \; u_2 = FT[u_2(x,z)] \; ; \; u_4 = FT [u_4(x,z)].$$

The backward wave generated through 4WM in the crystal, $u_3(x,z)$ evaluated at a distance f from lens L, is of the form :

$$u_3(x_0,z_0) = \alpha_0 u_1(-x,-z) . u_2(-x,-z) \times u_4(-x,-z),$$

where α_0 depends on the amplitude of the photoinduced index modulation;

. and **x** denote respectively the product and the convolution product of the optical field. Fig.23 illustrated another configuration based on 2WM in BSO. The interference pattern recorded in the Fourier plane of lens L is read out by an auxiliary low power laser (He-Ne ; λ_R = 633 nm).[61] The thickness of the crystal implies that this readout beam of different wavelength has to be positioned at the correct Bragg angle for obtaining the maximum intensity in the diffracted cross-correlation peak. Experimental results using BSO as a dynamic matched filter are shown in Fig.23 (typical response time τ = 50 ms for 1 mW incident intensity on the photorefractive crystal). The Bragg angular selectivity of the phase volume hologram may limit the number of pixels that can be processed and further studies should quantitatively estimate the capabilities of these architectures for high capacity parallel optical processing of analog or digital images. In these experiments, an auxiliary beam incident on the photorefractive crystal may also be used to modify the output of the processor in real time; this provides a means of weighting the correlation product in favor of specified spatial frequencies. Moreover, the introduction of a two-dimensional spatial light modulator (SLM) for real time data input allows demonstrations of a dynamic optical processor. Two-dimensional SLMs were demonstrated using optical erasure of photorefractive gratings.[62] Considering the technological progress on other input-output interfaces, these parallel processors should produce attractive new developments for application to pattern recognition and analog optical computing.

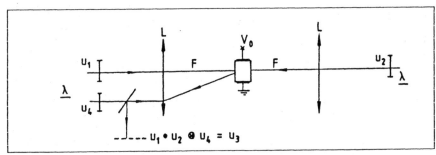

Fig.22: Application of 4WM in photorefractive crystals to image convolution and correlation. After (60).

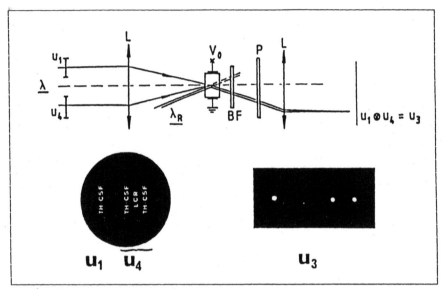

<u>Fig.23</u>: Application of 2WM to image convolution or correlation.
(BF) blocking filter for λ_R. Polarizer for noise filtering.
From (61).

5.
OPTICAL MEMORIES WITH PHOTOREFRACTIVE CRYSTALS

5.1. Introduction

Dynamic photorefractive crystals potentially offer two key features for application to parallel information storage. First, high storage capacity (diffraction limit $\lambda = 10^{12}$ bits/cm³) are available. Multiple data planes can be recorded and selectively retrieved using the Bragg selectivity of volume holograms. Secondly, information retrieval by association can be implemented when nonlinear photorefractive-based optical feedback is provided to the memory.

5.2. Multiple image storage

One of the peculiarity of a photorefractive crystal is its ability to store information in the dark. In absence of illumination, and at room temperature, the crystal dielectric relaxation time is simply given by $\tau_{di} = \varepsilon_0 \varepsilon / \sigma_0$ where σ_0 is the dark conductivity. Most of the ferroelectric crystals have a low conductivity in the dark $\sigma_0 < 10^{18}$ $\Omega^{-1}cm^{-1}$ and the related memory times range from 10 hours for KTN, to weeks for SBN, $BaTiO_3$ and several months for $LiNbO_3$. In photorefractive BSO-BGO, the measured dark crystal conductivity is about 10^{-14} $\Omega^{-1}cm^{-1}$ and the observed memory time is typically 10-20 hours. The dark storage time is considerably reduced in low bandgap semiconductor materials sensitive in the near IR and is typically 10^{-4} sec. in InP:Fe.

The superposition of many holograms in the same crystal volume is accomplished for example by varying the angle of incidence of the reference beam (Fig.24), each individual hologram being associated with a well-defined angle. Selective reconstruction of the superimposed holograms is achieved because efficient diffraction occurs only when the hologram is addressed at the right Bragg angle. This multiplexing technique requires a photorefractive crystal with an asymmetric recording erasure cycle (for example $LiNbO_3$ - 0.1% Fe^{3+}) in order to prevent erasure of previously recorded data pages during the recording of a new one in the same crystal volume. At present, up to 500 holograms have been superimposed and fixed in a cube of $LiNbO_3$ crystal giving a total capacity of 0.5 Gbits cm^{-3}.[63] A method for selective erasure of any information block, based on a coherent image subtraction technique, has also been demonstrated.[64]

The possibility of electrically controlled volume hologram writing in $LiNbO_3$ crystals has been demonstrated.[42] Due to the large electro-optic effect in this crystal, it is possible to control the Bragg conditions for image reconstruction by the bias voltage applied to the crystal. The independent reconstruction of two or three holograms written in the same crystal under different voltages is demonstrated in Ref.(42). The maximum efficiency is reached at the same voltage as was used for the recording.

5.3. Associative memories

Associative memory systems that use holographic data bases and phase conjugate mirrors to provide regenerative optical feedback, thresholding and gain have been recently reported.[65,66] The principles of information retrieval by association using parallel optical techniques, and in particular those based on holographic principles where recognized early by various authors.[67,68] However, these first approaches were limited in their ability to faithfully reconstruct the output object from a partial

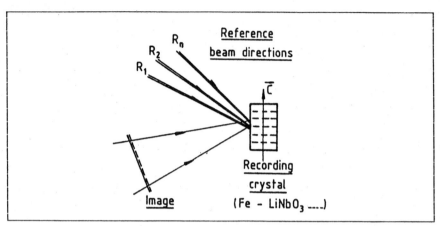

Fig.24: Multiple image storage in the volume of a photorefractive
crystal by angular coding of the reference beam direction.

input because of the large cross-talk which results when multiple objects
are holographically stored in the memory. Nonlinear elements such as
photorefractive crystals now permit these problems to be overcome, since
they introduce optical feedback and gain, thus improving the selectivity
and the stability of the memory. The principle of a holographic
associative memory is shown in Fig.25. Only a single hologram is used in
this configuration and it is simultaneously addressed by the object as
well as by the conjugate reference beam, the latter beam acting as the key
that unlocks the associated information. A photorefractive $BaTiO_3$ phase
conjugator is used both for reference beam retroreflection as well as for
gain and thresholding. This provides the necessary non linearity
emphasizing only the strongly correlated signals. The demonstration of
total image reconstruction of an object image when only a partial image
addressed the system, is also shown in Fig.25. The illumination of the
hologram by part of the object generates a diffracted beam propagating in
the original direction of the reference beam. This beam is then phase
conjugated and amplified by four-wave mixing in a photorefractive $BaTiO_3$
crystal. When this readout beam impinges on the hologram, it is diffracted
and recreates the initial object beam. This recreated object beam contains
all the information originally recorded in the hologram memory. This
principle can be extended to different objects superimposed in the
hologram memory by angular coding of the reference beam directions.[65,69]

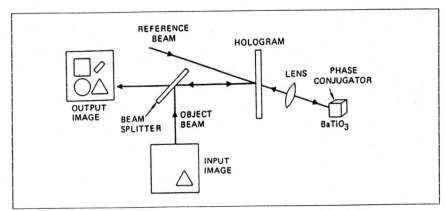

Fig.25: Associative holographic memory. Complete object image
reconstruction from a partial input image. From (65).

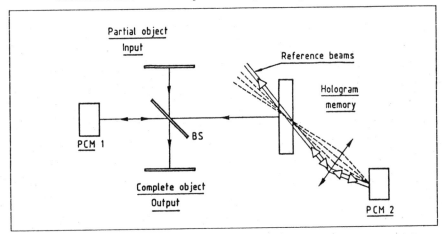

Fig.26: Optical associative memory using a phase conjugate resonator
and information storage in a volume holographic memory.
From (70).

A schematic implementation of such a nonlinear associative memory
based on a hologram placed in a cavity formed by two phase conjugate
mirrors is illustrated in Fig.26. The phase conjugate mirrors provide beam
retroreflection with gain and thresholding and give a self-alignment of
the object and reference wavefronts with respect to the hologram memory.
The optical feedback and the thresholding effects due to the non linear
mirrors favours the strongly correlated signals and forces the system to
converge to a stable state. The steady-state output signal thus consists
of the image stored in the holographic memory and that which presents the
highest degree of correlation with the input image. Real-time modification
of the memory is also possible if holograms are stored in the volume of a

photorefractive crystal. This would be a requirement for adaptive or learning behaviour of the system. This ability to reconstruct an image from a partial input information plane has important implications for pattern recognition, robotic vision and image processing operations. The nonlinear holographic associative memory also constitutes a first step towards the optical implementation of neural networks. This model is based on the feasability of distributed and interconnected memory elements with a nonlinear feedback, and is thus analogous in many aspects to optical holography.

6.

CONCLUSION

We have given a description of the photorefractive effect and its potential applications to some operations for use in the field of optical computing. Presently, the basic physical mechanisms leading to the photoinduced index in different crystals are well identified, but despite the great interest for applications, little is known about the charge transport processes or the species responsible for the photorefraction in these crystals. Therefore, continued research into the microscopic origins of the photorefractive effect is essential for further optimization of the materials through impurity doping and thermal treatments. It is hoped that crystals with much higher sensitivity and displaying larger photo-induced nonlinearities in the visible and near IR wavelength range will be developed in the future. Several of the existing materials already work at millisecond speed with low-power visible and near-IR lasers. Moreover, we have shown that the existence of a spatial phase shift between the incident fringe pattern and the photoinduced index modulation leads to the amplification of low intensity wavefronts through the energy transfer from a pump beam in two-wave or four-wave mixing interactions. The amplification factor depends on several recording parameters such as the grating period, the beam ratio of the interfering waves, the applied electric field as well as on the intrinsic material parameters. To summarize, photorefractive crystals certainly have most of the characteristics needed for initial demonstration of parallel optical processing devices. These capabilities will stimulate further studies on materials, algorithms and architectures for a variety of new applications using optical interactions in all-optical and hybrid systems.

REFERENCES

1. A.M.Glass, P.Gunter, J.P.Huignard, M.B.Klein, E.Kratzig,
   N.V.Kukhtarev, J.F.Lam, R.A.Mullen, M.P.Petrov, O.F.Schirmer,
   S.I.Stepanov, J.Strait, G.C.Valley in: "Photorefractive materials and
   their applications" Vol.61, Ed. by P.Gunter and J.P.Huignard
   (Springer-Verlag Berlin, 1988)

2. A.M.Glass: Opt. Eng. 17, 470 (1978)

3. P.Gunter: Phys. Rep. 93 (1982)

4. N.V.Kukhtarev, V.M.Markov, S.G.Odulov, M.S.Soskin, V.L.Vinetskii:
   Ferroelectrics 22, 949 (1979)

5. L.Young, W.K.Y.Wong, M.L.Thewalt, W.D. Cornish: Appl. Phys. Lett. 24,
   264 (1974)

6. F.Micheron: Ferroelectrics 18, 153 (1978)

7. G.C.Valley, M.B.Klein: Opt. Eng. 22, 704 (1983)

8. H.Kogelnik: Bell Syst. Tech. Jour. 48, 2909 (1969)

9. J.Feinberg, D.Heinman, A.R.Tanguay, R.W.Hellwarth: Jour. Appl. Phys.
   51, 1297 (1980)

10. R.R.Neurgaonkar, W.K.Cory, J.R.Oliver, M.D.Ewbank, W.F Hall: Opt. Eng.
    26, 292 (1987)

11. P.Gunter, F.Micheron: Ferroelectrics 18,27 (1978)

12. R.A.Mullen, R.W.Hellwarth: Jour. Appl. Phys. 58, 40 (1985)

13. A.M.Glass, A.M.Johnson, D.H.Olson, W.Simpson, A.A.Ballman: Appl. Phys.
    Lett. 44, 948 (1984)

14. M.B.Klein: Opt.Lett. 9, 350 (1984)

15. J.P.Herriau, D.Rojas, J.P.Huignard, J.M.Bassat, J.C.Launet:
    Ferroelectrics 75, 271 (1987)

16. H.Rajbenbach, J.P.Huignard, B.Loiseaux: Opt. Commun.48, 247 (1983)

17. B.Imbert, H.Rajbenbach, S.Mallick, J.P.Herriau, J.P.Huignard: Opt.
    Lett. 13, 327 (1988)

18. V.Kondilenko, V.Markov, S.Odulov, M.Soskin: Optica Acta 26, 238 (1979)

19. A.Marrakchi, J.P.Huignard, P.Gunter: Appl. Phys. 24, 131 (1981)

20. J.P.Huignard, A.Marrakchi: Opt. Commun. 38, 249 (1981)

21. R.W.Hellwarth: Jour. Opt. Soc. Am. 67, 1 (1977)

22. J.P.Huignard, J.P.Herriau: Appl. Opt. 24, 4285 (1985)

23. G.C.Valley: Jour. Opt. Soc. Am. B1, 868 (1984)

24. Ph.Refregier, L.Solymar, H.Rajbenbach, J.P.Huignard : Jour. Appl. Phys. 58, 45 (1985)

25. G. Hamel de Monchenaux, B.Loiseaux, J.P.Huignard: Appl. Phys. Lett. 50, 1794 (1988)

26. G. Hamel de Monchenaux, J.P.Huignard: Jour. Appl. Phys. 63, 624 (1988)

27. R.B.Bylsma, A.M.Glass, D.H.Olson: Elec. Lett. 24, 362 (1988)

28. D.Rytz, M.Klein, R.A.Mullen, R.N.Schwartz, G.C.Valley, B.A.Wechsler: Appl. Phys. Lett. 52, 1759 (1988)

29. V.Markov, S.Odulov, M.Soskin: Opt.Laser Tech.11, 95 (1979)

30. F.Laeri, T.Tschudi, J.Albers: Opt. Comm. 47, 387 (1983)

31. Y.Fainman, E.Klancnik, S.H.Lee: Opt. Eng. 25, 228 (1986)

32. J. Feinberg in: Optical phase conjugation, Ed. by R.A.Fisher (Academic Press, London, 1983)

33. J.P.Huignard, J.P.Herriau, Ph.Aubourg, E.Spitz: Opt. Lett.4, 21 (1979)

34. S.Odulov, M.Soskin, V.Vasuetsov: Opt. Commun. 32, 183 (1980)

35. P.Gunter: Opt. Lett. 7, 10 (1982)

36. J.Feinberg, R.W.Hellwarth: Opt. Lett. 5, 519 (1980)

37. B.Fischer, M.Cronin-Golomb, J.O.White, A.Yariv, R.Neurgaonkar: Appl. Phys. Lett. 40, 863 (1982)

38. H.Rajbenbach, J.P.Huignard, Ph.Refregier: Opt. Lett. 9, 558 (1984)

39. H.Rajbenbach, B.Imbert, J.P.Huignard: Submitted to Opt. Lett. (1988)

40. S.I.Stepanov, M.P.Petrov: Opt. Acta 31, 1335 (1984)

41. D.Rak, I.Ledoux, J.P.Huignard: Opt. Commun. 49, 302 (1984)

42. M.P.Petrov, S.I.Stepanov, A.A.Kamshilin: Opt. Commun. 21, 297 (1977)

43. G.Pauliat, J.P.Herriau, A.Delboulbé, G.Roosen, J.P.Huignard: Jour. Opt. Soc. Am. B3, 306 (1986)

44. J.W.Goodman, F.J.Leonberger, S.Y.Kung, R.A.Athale: Proc.IEEE 72, 850 (1984)

45. P.D.Henshaw: Appl.Opt. 21, 2323 (1984)

46. M.Cronin-Golomb, B.Fischer, J.O.White, A.Yariv: IEEE QE 20, 12 (1984)

47. H.Rajbenbach, J.P.Huignard: Opt.Lett. 10, 137 (1985)

48. J.P.Huignard, H.Rajbenbach, Ph.Refregier, L.Solymar: Opt. Eng. 24, 586 (1985)

49. A.Yariv, S.K.Kwong: Opt. Lett. 10, 454 (1985)

50. M.D.Ewbank, P.Yeh: Opt. Lett. 10, 496 (1985)

51. H.Rajbenbach, Y.Fainman, S.H.Lee: Appl. Opt. 26, 1024 (1987)

52. J.Feinberg, R.W.Hellwarth: Opt. Lett. 5, 519 (1982)

53. Y.Fainman, C.Guest, S.H.Lee: Appl. Opt. 25, 1598 (1986)

54. H.Rajbenbach: Jour. Appl. Phys. 62, 4675 (1987)

55. J.Feinberg: Opt. Lett. 7, 486 (1982)

56. G.A.Rakuljic, K.Sayano, A.Yariv, R.R.Neurgaonkar: Appl. Phys. Lett. 50, 10 (1987)

57. S.K.Kwong, G.A.Rakuljic, A.Yariv: Appl. Phys. Lett. 48, 201 (1986)

58. A.E.Chiou, P.Yeh: Opt. Lett. 11, 306 (1986)

59. D.Z.Anderson, M.C.Erie: Opt. Eng. 26, 435 (1987)

60. J.White, A.Yariv: Appl. Phys. Lett. 37, 5 (1980)

61. L.Pichon, J.P.Huignard: Opt. Commun. 36, 277 (1981)

62. A.Marrakchi, A.R.Tanguay Jr., J.Yu, D.Psaltis: Opt. Eng. 24, 124 (1985)

63. D.L.Staebler, W.Burke, W.Philips, J.J.Amodei: Appl. Phys. Lett. 26, 182 (1975)

64. J.P.Huignard, J.P.Herriau, F.Micheron: Appl. Phys. Lett. 26, 256, (1975)

65. B.H.Soffer, G.J.Dunning, Y.Owechko, E.Marom: Opt. Lett. 11, 118 (1986)

66. D.Z.Anderson: Opt. Lett. 11, 56 (1986)

67. R.J.Collier, K.S.Pennington: Appl. Phys. Lett. 8, 44 (1966)

68. D.Gabor: IBM Jour. Res. Develop. 13, 156 (1969)

69. A.Yariv, S.Kwong, K.Kyuma: Appl. Phys. Lett.48, 114 (1986)

70. B.Fischer, S.Sternklar, S.Weiss: Appl. Phys. Lett. 48, 1567 (1986)

DIGITAL OPTICS

Adolf W. Lohmann
University of Erlangen, FRG

1.

INTRODUCTION

The term "digital optics" needs to be defined, which is the aim of chapter 2. Thereafter we will present a short history of our topic (chapter 3). Microelectronics may be seen as a competitive technology. That is true, to some degree. But we view digital optics and microelectronics more as complementary approaches towards a common goal: information processing. This point requires some clarification, which will be dealt with in chapter 4, called "motivation". The remaining chapters demonstrate how classical optics can be harnessed to solve some specific problems in digital optics.

Other areas of digital optics will be deferred to other lectures within this series, for example nonlinear devices to D.A.B. Miller and architecture to K.-H. Brenner, among others.

2.

DIGITAL OPTICS - A NEW TECHNOLOGY FOR COMPUTERS

Today optical data communication through fibers is in general use. Optical mass storage has become a common technology. For data processing, however, the optical signals have to be converted into electrical signals, so far. Today, all-optical processing is a challenging research problem.

Digital Optics is a comprehensive optical technology for (i) digital optical processing, (ii) data transport and (iii) information storage.

The key components for digital optical data processing are (i) two-dimensional opto-electronical or opto-optical logic gate arrays, (ii) modules for parallel optical interconnections through free space and for beam shaping and (iii) system design and architectures adapted to optics.

The key technologies are (i) semiconductor materials and processing for the optical logic gate arrays, (ii) classical optics, micro-optics and holography for interconnections and beam shaping and (iii) laser technology for the optical power supply of the gate arrays.

Digital Optics is different from integrated optics and from fiber communications: waveguides are good for long-distance interconnections at high data rates with low attenuation. Since waveguides are typically point-to-point interconnects it is cumbersome to connect more than a few participants or terminals. On the other hand, in a data processing system we find mainly moderate- and short-distance interconnections. A large number of these interconnections is necessary in complex digital circuits. Free-space optical interconnects by imaging allow parallel access to two-dimensional arrays of data.

The connectivity (i.e. the number of independent data channels per chip) is 10^4-10^6 in optics. Electronic integrated circuits have a connectivity on the order of a few hundred. By the way, the large connectivity of optics is illustrated also by the high information density in optical data storage.

Hence, in the future high performance digital circuits will make use
of optical interconnections because of two main reasons: (i) the large
temporal bandwidth of optics and (ii) the large connectivity of optics
that allows parallel communications. The energy for optical switching and
logic is nowadays comparable with electronics.

3.
BRIEF HISTORY OF DIGITAL OPTICS

A brief table indicates the main events in the history of optical
information   processing.   There are probably   several   pre-historic
forerunners. Stonehenge might be among them. John von Neumann, the
ingenious inventor of the programmable computer and of its communications
bottleneck, contemplated very early about the use of electromagnetic
waves as data carriers instead of electrons. Analog procesors were
already in use, when the laser triggered the first experiments in digital
optics around 1960. Ten years later a systematic review of optical
effects as a base for an "optical transistor" turned out to be quite
pessimistic. This conclusion was regrettable, because some optimism about
the usefulness of uncommon number systems as a base for optical computing
emerged about that time.

The movement of digital optics gained momentum, when optical
bistability became established and when novel parallel architectures were
developed especially of implementation by optical means.

Time table of digital optics
~ 1945 : B.C.(?) Stonehenge, a binary clock?
~ 1945 : John von Neumann files a patent:
         Computer, Running on El. Magn. Waves
~ 1955 : Optical Analog Computers:
         Side-Looking Radar,
         Acousto-Optical Spectrum Analyzer
~ 1960 : Optical Nor-Gate by means of Laser Quenching;
         Optical Parallel Architecture based on Spatial
         Filtering ("Theta modulation"),

~ 1970 : Optical Nonlinearities too weak, shown by a
         systematic survey.
         Residue numbers good for optics?
         Cyclic effects, such as interference, suitable
         for residue conversion.
~ 1980 : New optical effects: bistability, amplification by
         feedback
         New parallel architectures: symbolic substitution,
         OPALS, systolic arrays, associative memory, binary
         image algebra, cellular arrays, etc.

4.

MOTIVATION

The progress of microelectronics is impressive. Nevertheless, the rate of
price/performance seems to level off, while the demand for more computer
power is steadily increasing. There is not very much to be gained anymore
in terms of the hardware speed. GaAs may be faster than Si by one order
of magnitude. Hence, the only hope rests on coupling many processors
together in parallel.

Architecture:

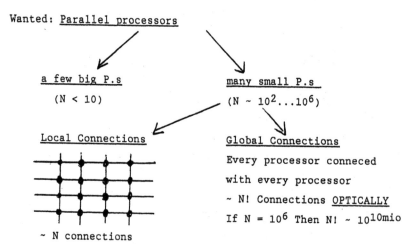

Wanted: Parallel processors

a few big P.s   many small P.s
  (N < 10)   (N ~ 10^2...10^6)

Local Connections   Global Connections
                                 Every processor conneced
                                 with every processor
                                 ~ N! Connections OPTICALLY
                                 If N = 10^6 Then N! ~ 10^{10}mio
~ N connections

The prices/performance ratio will probably favor many small processors instead of a few big processors.

The next decision branch relates to how these many small processors are going to be connected. The variety of communication networks is large. The cartesian grid is simple but slow, if data are to be moved from one end to another. The other extreme would consist of $(N-1)!$ permanent lines between N processors. A flexible crossbar net with N^2 binary switches may be desirable. But it is still too expensive if N is reasonably large. We will consider later on other networks with $O[N \log N]$ switches.

One may wonder if global connections are really needed. For example, every processing element may represent one point in (x, y, z) space. Partial differential equations describe short distance forces. Hence, a 3D processor grid ought to be quite suitable for implementing an algorithm in isomorphic fashion. An important counter example is the Fourier transform, which requires global data paths.

Communication, as the crucial aspect of high performance data processing, is evident on every level of a computer system. The relative increase of cost for communication in comparison with cost for switching is shown in the following chart.

Relative Costs

Switching	Cost	Communication	t
Tubes	>>>	Wires	1950
Transistors	>>	Wires	1960
IC-Transistors	≈	Planar Guides	1970
Ic-Transistors	«	Multi Layer Guides	1980

Multilayer chips are an expensive way of overcoming the communication difficulties in single-layer planar chips.

The cause of the communications problems can be inferred from the next chart, which starts from the simplified assumption: two things occur in a computer: logic interactions and transport of signals. Electrons do interact, which is good for logic but disturbing for communication. Photons normally do not interact, which is good for communication. But it makes logic impossible. Fortunately, photons can be enticed to interact if they meet within a piece of exotic material. Such material will be placed only where interactions are wanted.

ELECTRONS OR PHOTONS ?

LOGIC INTERACTIONS	+ + +	−	+
TRANSPORT COMMUNICATION	+ +	+ + +	+ +

If we assign grades to electrons and photons for the performance in
switching and in communication the following three-step scenario emerges:

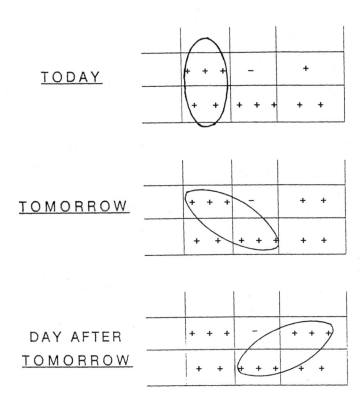

These charts served to indicate the fundamental reasons for our
optimism. The consequences, drawn by various research groups, point into
at least six different directions. Direction 5 will be emphasized here.

Directions of R&D in Digital Optics

(1)   Subsystems: Rom, Worm, Bus, I/O

(2)   Hardened   : EMI, EMP

(3)   Very fast switches: 10^{-12}sec

(4)   Binary logic by polarisation
      instead of by "on-off" power

(5)   Massively parallel
      based on communication in free space

(6)   Neural processing
      massively parallel threshold logic

5.

FREE-SPACE OPTICAL WIRING, FOR EXAMPLE CYCLIC SHIFTING

It is quite simple in optics to shift an image laterally (compare parts
(a) and (b) in the figure below). One merely inserts a prism or a glass
wedge somewhere between the object and image plane. For data processing
one needs sometimes a modified shift, called a "cyclic shift". Whatever
is pushed beyond the edge of the frame will re-appear on the other end of
the frame (see part (c) of the figure below).

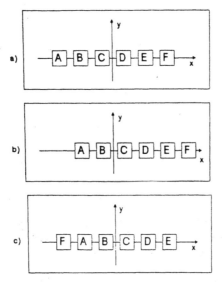

One   particular   way of implementing a cyclic shift is   shown   in   the
following figure (from ref. 1).   This example serves as illustration   for
the simplicity of optical parallel wiring in free space.

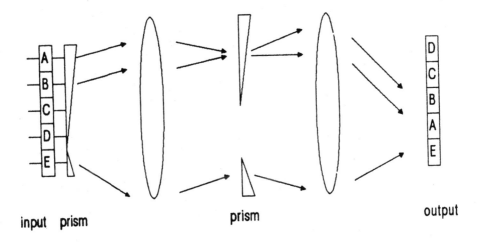

input prism   prism   output

6.

THE PERFECT SHUFFLE

Probably more important than cyclic shifting is another job of channel
arrangement, called "perfect shuffling". The name stems from card
playing, where a perfect mixing means alternate interlacing (see figure
below). The bit planes of the binary address numbers of the channels are
shifted cyclincally. The mask (right hand side of the figure) can be
replaced by lossless holographic telescope arrays (ref. 2).

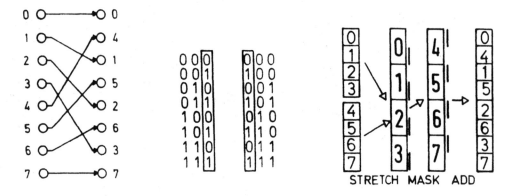

THE "PERFECT SHUFFLE"

The perfect shuffle in its electronic version is in use already
(ref. 3-6) in the communications industry. Several optical implementations of the perfect shuffle have been suggested and studied (ref 7-16).
A two-dimensional version (ref. 8) with two lenses and four prisms is
shown in the figure below.

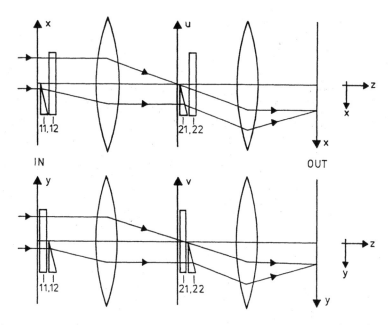

PS FOR TWO-DIMENSIONAL DATA ARRAYS

7.

ARRAY ILLUMINATORS

Uniform arrays of optical gates will have to be illuminated for various reasons, such as optical power supply, holding power of amplifiers or bistable elements, or readout illumination. To that end the power of a big laser (say 1 Watt = 10^{10} photons per second) has to be split into up to $(1000)^2$ beamlets. This amounts to one micro Watt per channel. If each channel may operate at nanosecond speed, then there are only a thousand photons per switching act. It is obvious, in view of these numbers, that any waste of photons ought to be avoided. Furthermore, the homogeneity in space and in time ought to be maintained within a reasonable tolerance limit.

We have designed four different approaches for the illumination of arrays, as shown in the figure below.

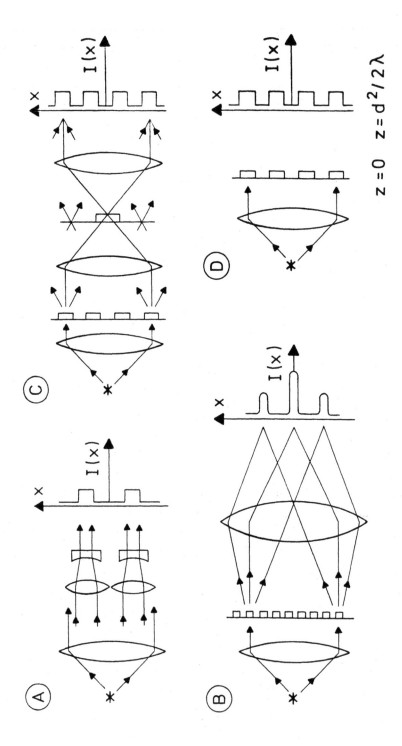

The underlying concepts are arrays of holographic lenslets (fig. A, ref. 17, 18); grating diffraction (fig. B, ref. 19-22); phase contrast (fig. C, ref. 23) and Talbot self-imaging, based on Fresnel diffraction (fig. D, ref. 24).

Many co-workers have contributed to the project "Digital Optics", of a small fraction. I am grateful for many lively interactions and for dedicated work. These co-workers are (in alphabetic order): H. Bartelt, K.-H. Brenner, L. Frank, J. Jahns, G. Lohman, T. Merklein, F. Sauer, J. Schwider, W. Stork, N. Streibl, G. Stucke, J.A. Thomas, R. Völkel, J. Weigelt and a few others.

References

1. K.-H. Brenner, A.W. Lohmann, Appl. Opt. $\underline{27}$ (1988) 434.

2. A.W. Lohmann, F. Sauer, Appl. Opt. $\underline{27}$ (1988) 3003.

3. H.S. Stone, IEEE, $\underline{C20}$ (1971) 153.

4. C.L. Wu, T.Y. Feng, IEEE Trans.Comp. $\underline{c-30}$ (1981) 324.

5. M.E. Marhic, Opt. Lett $\underline{9}$ (1984) 308.

6. C.J.G. Kirkby, Electron. Lett. $\underline{23}$ (1987) 971.

7. A.W. Lohmann, W. Stork, G. Stucke,
   Appl Opt. $\underline{25}$ (1986) 1530.

8. A. W. Lohmann , Appl. Opt. $\underline{25}$ (1986) 1543.

9. K.-H. Brenner, A. Huang, Appl. Opt. $\underline{26}$ (1987) 135

10. S.-H. Linm, T.F. Krile, J.F. Walkup,
    SPIE Proc. $\underline{752}$ (1987) 209; Appl. Opt. $\underline{27}$ (1988) 1734.

11. G. Eichmann, Y. Li, Appl. Opt. $\underline{26}$ (1987) 1167.

12. G. Stucke, Optic $\underline{78}$ (1988) 84.

13. C.W. Stirk, R.A. Athale, M.W. Haney,
    Appl. Opt. $\underline{27}$ (1988) 202.

14. Q.E. Song, F.T.S. Yu, Appl. Opt. $\underline{27}$ (1988) 1222.

15. K.M. Johnson, M.R. Surette, J. Shamir
    Appl. Opt. $\underline{27}$ (1988) 1727.

16. G.E. Lohman, A. Lohmann, Opt. Eng. $\underline{27}$ (1988) Oct.

17. H. Bartelt, N. Streibl, Opt. Eng. $\underline{24}$ (1985) 1038.

18. A.W. Lohmann, F. Sauer, Appl. Opt. $\underline{27}$ (1988) 3003.

19. G. Groh, Appl. Opt. 7 (1988) 1643; 8 (1969) 967.

20. H. Dammann, K. Görtler, Opt. Commm. 3, 312 (1971).

21. H. Dammann, E. Klotz, Opt. Acta 24 (1977) 505.

22. M. Prise, N. Streibl, M. Downs,
     Optical and Quantum Electronics 20, 49, (1988).

23. A. W. Lohmann, J. Schwider, N. Streibl, J.A. Thomas,
     Appl Opt. 27 (1988) 2915.

24. A.W. Lohmann, Optik 79, 41, (1988).

OPTICAL INTERCONNECTIONS

Joseph W. Goodman
Department of Electrical Engineering
Stanford University
Stanford, California 94305 U.S.A.

Background

In most basic terms, a digital computing structure consists of two different types of elements, both necessary for the system to function. On the one hand we have nonlinear gates within which signals must interact with each other. Accompanying these elements we have interconnections between gates and between groups of gates of various sizes. The interconnections provide the means of communication between gates.

There is a wide variety of interconnect problems within a computer. Interconnects can be "high-level" or "low-level", depending on the level of the architecture at which they function. The following list discusses various types of interconnects, proceeding from high level to low level.

Machine-to-machine interconnections are the most mature of the various optical interconnect technologies. We refer to fiber-optic local-area networks, which may interconnect as many as hundreds of different computers over distances of several km. Commercial products already exist, and will become more plentiful as the new optical LAN standard, named FDDI, is fully specified. FDDI will be a 100 Mb/s system utilizing multimode fiber and LED/PIN diode terminals.

Backplane-to-backplane interconnections are the next likely candidate for optics. A backplane is a motherboard into which the various electronic boards comprising an electronic system plug. The backplane must supply communication capability between boards. With a optical backplane-to-backplane extender, boards in one backplane can, for example, address the memory of boards in another distant backplane. Optical backplane-to-backplane interconnections have been used in the AT&T ESS-5 switching computer, and a VMEbus backplane-to-backplane interconnect will soon be available commercially.

Board-to-board interconnections are provided by the backplane. Thus an optical backplane would replace the multilayer electronic motherboard commonly used today. Total data rates should be in the multiple GHz range.

Chip-to-chip interconnections exist at the board level. They are currently carried out by means of conducting lines in a layer or layers of the board itself. Optics could provide the communication paths between chips in several different ways, using optical fibers, integrated optic waveguides, or holographic optical elements.

Gate-to-gate interconnections are required at the chip or wafer level. A single integrated circuit chip typically contains thousands of such interconnections. Optics has been considered for use at this, the lowest level of the interconnect hierarchy.

One of the open questions in this branch of optical computing is how far down the hierarchy of interconnections optics can penetrate. Penetration has already been made at the highest level, machine-to-machine communication. For reasons of power dissipation, we will argue later that it is unlikely that optics will penetrate the lowest level of the interconnect hierarchy, gate-to-gate interconnects. Thus at some intermediate level, optics may not be a viable approach, at least with today's technology. Just where that level lies is a question of some interest.

Properties of Optical Interconnections

Optical signals have certain properties that make them very attractive for solving interconnect problems. Interconnections should behave in a linear fashion and should be as free as possible from crosstalk generated by other adjacent interconnections. Electrons and photons have certain fundamentally different properties in this regard. Since electrons are charged particles, they have a propensity to interact with other nearby electrons, as evidenced by the fundamental properties of Maxwell's equations. These interactions become stronger and stronger as the bandwidth of the interconnect signals grows. On the other hand, photons, being uncharged, do not exert an influence on one another in any linear medium, and indeed can pass through one another without any interaction taking place. Thus optical interconnections offer the promise of a fundamental freedom from mutual interference, although in practice there are always practical effects, such as scattering, that induce some level of crosstalk.

A second advantage of optical interconnections lies in the comparative simplicity with which optical signals can be terminated. Electrical lines require complex impedance matching terminations in order to prevent reflections at the end of the line. Such terminations become harder and harder to make as the bandwidth of the signals goes up. In addition, as fan out, or tapping of the signal, occurs down an electrical line, impedance discontinuities are generated and reflections result. For optical beams, no such phenomenon exists. Backreflections can be suppressed in a multitude of ways, using index-matching terminators, for example. Since the modulation bandwidth is but a tiny fraction of the optical center frequency, termination becomes quite simple.

A third advantage of optics lies in its freedom from planar or quasi-planar constraints. Electrical interconnections must be placed near a ground plane, which provides a termination for electric field lines that would otherwise terminate on adjacent interconnect lines, causing crosstalk. On the other hand, optical signals are well confined spatially and can be routed quite flexibly through three-dimensional space.

With these motivations in mind, we turn to a consideration of the more detailed nature of optical interconnects.

Forms of Optical Interconnections

There are several in which an optical interconnection can be realized. The particular form used in practice may depend on the level of the interconnect hierarchy of interest.

The options available can be summarized as follows: 1) fiber optics, 2) free-space optics, or 3) waveguides in substrates. The use of fiber optics for establishing interconnections is particularly appropriate at the high levels of the hierarchy. Free-space optics have been used to interconnect multiple processors in a single machine [1] and have been studied for chip-to-chip communication on a board [2]. Waveguides on substrates are most appropriate at the chip-to-chip level and possibly for clock distribution on a single chip [3].

In the free-space category are holographic optical elements for interconnects, which function to image a set of sources onto a set of detectors, with the greatest efficiency possible.

If the optical channels are to run at high speed, then fast photodetectors are needed, which in turn generally implies that the photodetectors are quite small. Thus whatever technology is used must be capable of delivering light efficiently to a small photodetector at the receiving end.

Efficiency of Optical Interconnects

The ultimate purpose of an interconnect, whether electrical or optical, is to charge a gate capacitance to the threshold voltage at the end of the interconnect line. Use of an optical interconnect implies conversion of electrons to photons and back to electrons. Thus the efficiency of these conversion processes is critical, as is likewise the efficiency with which light is delivered from source to detector.

If the interconnect is realized at 800 nm wavelength, a PIN photodetector can be expected to have an 80% quantum efficiency or a responsivity of about 0.5 Amps/Watt. Thus four of every five received photons is converted into an electron. Avalanche photodiodes provide more efficient conversion, typically having responsivities in the 50 to 100 Amps/Watt range, but at the price of excess noise, as well as high voltages required for operation. The corresponding numbers at other optical communication wavelengths are not markedly different. In general it can be said that the conversion of photons to electrons is a relatively efficient process.

As for the optical source, light-emitting diodes (LED's) typically have efficiencies in the 1% to 2% range, particularly when low coupling efficiencies are included, and therefore are comparatively unattractive in terms of electrical power required to generate a given amount of optical power. In addition, their modulation rates are typically limited to the range 100-200 MB/s. On the other hand, LED's are very reliable and require no thermal control, and hence are often the source of choice at high levels of the interconnect hierarchy, provided the losses in the communication path are not too large. More efficient electrical -to-optical power conversion can be obtained with laser diodes. While high-power laser diodes have been demonstrated with 50% conversion efficiency, communication-type diodes, with outputs of about 1 mW of power, typically are limited to overall efficiencies of 25% or less. These efficiencies are achieved only when the diode is driven above the threshold for lasing. Threshold currents vary widely, and of course the lower the threshold the better from the perspective of optical interconnects. Recently lasers have been reported with sub mA threshold currents (see, for example Ref. [4]). Such lasers have a minimum required electrical drive power of a very few milliwatts. It is important to note that if an optical interconnect using a laser diode is to be used, the minimum power commitment is of this order of magnitude. Such powers may not be competitive with electronic solutions at the lowest levels of the interconnect hierarchy (gate to gate), but may be quite advantageous at the higher levels of the hierarchy.

The efficiency with which light is transferred from the source to the detector is also important. Over the short distances of interest here, the losses associated with a fiber interconnect are quite small (a few dB at most, unless lossy elements such as switches are inserted). Free-space broadcast techniques can suffer from large losses. For example the uniform broadcast of a signal to a 1 cm x 1 cm chip, within which are imbedded a series of 10 μm x 10 μm detectors, can yield a loss of 60 dB. Holographic optical elements alleviate this difficulty by providing some focusing power. The holograms themselves are not perfectly efficient, typically introducing 10 dB to 3 dB of loss for silver-halide materials, and less for dichromated gelatin materials.

Power Comparisons for Optical and Electronic Interconnects

A power comparison has been made for a specific set of interconnect problems [5]. At the gate-to-gate level the electronic parameters are based on silicon technology. The data rate assumed is 1 Gb/s, and the threshold voltage was taken to be 1 volt. Linewidths of 0.5 μm were assumed and a line length of 1 mm was considered. The average power required to establish the interconnect was found to be about 35 μW. The corresponding optical solution , assuming a 25 μm x 25 μm detector with 0.5 Amps/Watt detector responsivity and a laser diode with an efficiency of 25% yields a minimum electrical drive power of about 1 mW. In fact, a 2 or 3 mW would be needed due to the finite threshold of the laser. Thus at the gate-to-gate level the optical solution is not competitive with the electronic solution.

At the chip-to-chip level, the numbers required of the electrical interconnect are different, while the optical interconnect numbers remain unchanged. The dominant power requirement in electrically interconnecting two chips over a distance of a few cm (assumed short enough to avoid the use of a terminated transmission line) is the reactive power required to charge the bonding pad capacitances. Again assuming a 1 Gb/s data rate and a 1 volt logic threshold, the power required is about 175 μW. On the other hand, if the distance is long enough (and the data rate high enough) to require the use of a terminated transmission line, the power dissipated in the terminating resistor will be about 10 mW. The optical power required is again 2 or 3 mW. Thus we see that from the power point-of-view, optical interconnects look attractive whenever the electrical solution requires a terminated transmission line.

A related conclusion has been reached by D.A.B. Miller [6]. The power required for an electrical interconnect is inversely proportional to the terminating impedance, which must be matched to the characteristic impedance of the line. The problem arises from the difficulty of realizing electrical transmission lines with large characteristic impedance. For example, a coaxial cable has a characteristic impedance that increases only logarithmically with the ratio of the diameters of the two conductors, making it practically very difficult to realize a high impedance. Miller points out that the optical interconnect serves as a type of impedance transformer, allowing the equivalent of a high characteristic impedance to be realized.

It should be remembered that drive power is just one characteristic of interest when comparing two interconnect technologies. It may be that the superior immunity of optics to crosstalk would justify its use in some situations where the power required exceeded that of an electronic equivalent.

Prognosis for the Future

Based on the discussion above, it seems likely that optics will penetrate the interconnect hierarchy in computing down to the level where the lines do not require termination. Beyond this point, electrical interconnects will probably be hard to compete with. Thus we can expect to see optics applied to problems of module-to-module communication, backplane-to-backplane communication, board-to-board communication (the optical backplane), and to some chip-to-chip communication problems. It is highly unlikely that optics will be useful at the gate-to-gate interconnect level.

References

[1] H. Tajima, Y. Okada, K. Tamura, "A high-speed optical common bus for a multiprocessor system", *Trans. Inst. Electron. and Commun. Eng. Japan*, Vol. 24, No. 17, pp. 850-866 (1984).

[2] R.K. Kostuk and J.W. Goodman, "Optical imaging applied to microelectronic chip-to-chip interconnects", *Applied Optics*, Vol. 24, No. 17, pp. 2851-2858 (1985).

[3] B.D. Clymer and J.W. Goodman, "Optical clock distribution to silicon chips", *Optical Engineering*, Vol. 25, No. 10, pp. 1103-1108 (1986).

[4] K.Y. Lau, N. Bar-Chaim, P.L. Derry, and A. Yariv, "High-speed digital modulation of ultralow threshold (<1 mA) GaAs single quantum well lasers without bias", *Appl. Phys. Let.*, Vol. 51, No. 2, pp. 69-71 (1987).

[5] J. W. Goodman, "Optics as an Interconnect Technology", in *Optical Processing and Computing* , H.H. Arsenault, Editor, Academic Press (in Press).

[6] D.A.B. Miller, "Low-energy optical communication inside digital processors: quantum detectors, sources and modulators as efficient impedance converters", Paper MT1, OSA Annual Meeting, Santa Clara, California, Oct. 31-Nov. 4 (1988).

OPTICAL SWITCHING NETWORKS FOR COMMUNICATION SYSTEMS

G Parry, D R Selviah and J E Midwinter
Department of Electronic and Electrical Engineering
University College London
Torrington Place
London, WC1E 7JE

1.

INTRODUCTION

Progress in fibre optic communication technologies and systems during the past decade has been extremely rapid with new ideas and technologies having a major impact on the range of services which telecomminications companies can now offer customers. While the progress has been impressive it must be admitted that the problem which systems designers have addressed has been very clearly defined. The problem has been simply to transmit data from point A to point B with minimum distortion and over a maximum range. This has not meant that the technological developments required have been easy, but with such a clearly defined problem the implications of any new technological development have become apparent very quickly and the economics can be argued forcefully. Transmission systems are now so successful that many telecommunications organisations are directing their interest to other problems and are asking

what impact opto-electronic technology can have on the solution
of those problems. The switching of very wide-band signals is
one topic which, most organisations agree, will become
important in the digital networks being considered for
combined transmission of a range of different telecommunication
services. The impact of optical switching in such a network is
one of the topics which we will consider in this lecture. We
will also consider more generally, how optical switching can
benefit other, simpler, networks.

Approaches to optical switching vary from all-optical
switching to opto-electronic switching, in which optical
signals are simply converted to electronic signals for
switching and reconverted to optical signals after switching.
Opto-electronic switching is used at present so we can be
confident that other entirely optical switching techniques will
only be implemented if they perform a function which is
impossible in opto-electronics, or if the purely optical
technique is cheaper or leads to a lower systems cost. This
sets some bounds on what optical technology must achieve.
Electronic switching at a few Gbits/s and in times of tens to
hundreds of picoseconds is actually within the capability of
advanced electronic technology, at least for relatively simple
systems. Interestingly the multi-Gbits/s data rate is also the
data rate which the next decade's switching systems will be
aiming at, so optical switching will have to take on
electronic switching and offer a more attractive solution. An
alternative strategy which is being adopted by some groups is
to aim to introduce all-optical switching into far higher data
rate systems which may be required on a much longer time scale.
In this lecture we will concentrate mainly on the shorter term
requirements.

Since the lecture is intended to be tutorial we will include an introduction to switching systems and some background material on optical communications. These will provide a view of the environment in which any optical switching systems must work.

2.
NETWORKS AND SWITCHING REQUIREMENTS

A communications network is an arrangement of transmission lines interconnected via switching nodes which allows the communication of information from any one input port to one or more selected receiver ports. The public switched telephone network (PSTN) is the most well known example. There are many other examples: local area networks (LANs), wide area networks (WANs), metropolitan area networks (MANs) all enable communication between a smaller number of users than the PSTN but which may offer higher bandwidth transmission systems for data transmission or video conference for example.

The development of communication networks has involved developments in switching technologies, transmission systems and in the design of switched circuits. Three types of switched circuit can be identified: circuit switching, message switching and packet switching. In circuit switching, an end to end link is formed by the exchange and the connection is held for the duration of the call. This permits interactive communication but it can make inefficient use of the network for data transmission. Message switching is based on the principle that information to be communicated between A and B can be sent via messengers who receive the information and either pass it straight on or, if the communication lines needed are busy, store the information until a suitable route can be identified. Message switching is clearly not interactive and can be very slow. Packet switching also involves the "store and forward" principle but it can be interactive. A message to be transmitted is divided into a number of short lengths known as packets. The packet consists, essentially, of a header to identify the address to which the packet is to be sent, a part of the message, and an error code, as shown below.

```
..............................................
. ERROR .      MESSAGE      . HEADER  .
..............................................
```

Fig. 1

A packet assembler-disassembler (PAD) is required at each terminal to prepare the message in the correct format or to reform the received packets into the original message. When packet data is sent from a transmitting port to a packet switch, the switch must be able to read the address and then route the packet to the correct receiving port. The transmission lines between different packet switches may carry a number of packets associated with different messages, interleaved to ensure efficiency and rapid message transmission. By breaking up the complete message in this way, sufficiently rapid communication can occur to permit interactive operation of the network. (Practical implementations of packet networks are considerably more complex than the simple picture given above.)

At the present time many of the networks in use are dedicated to one or two specific services. There is considerable current interest in developing a single network which could carry and switch a much wider range of services. This type of network is known as an integrated services digital network (ISDN) and it could carry video, facsimile, computer data, mail and voice information. The services to be provided by an ISDN impose a wide range of requirements on band-width, message lengths, call set up times etc. e.g. message lengths can vary from of a few 100 bits in length to data file transfer involving many Gbits and at rates from 64 kbit/s for telephony to 100-600 Mbits/s for high definition television. The packet network is one which is being considered for such applications[1]. It seems likely that if optical switching is to play a major role in future communications networks then optical switches will need to cope with the demands of an ISDN.

3.
SOME POSSIBLE ROLES FOR OPTICAL SWITCHING

We can identify a number of applications where optical
switching may play a key role:

- Routing of wide-band digital data in packet format in wide-
  band local networks to provide services to business or
  homes.

- Routing of wide-band digital data in trunk networks.

- Reconfiguration of parts of networks to correct faults or
  to cope with an overload.

- Switching wide-band video networks for security or video
  conference applications.

- Optical clock regeneration

- Multiplexing (time) and demultiplexing at high data rate
  (1-10Gbit/s)

4.
IMPLICATIONS FOR OPTICAL SWITCHING SYSTEMS

The major consequence imposed by network considerations is the
fact that for most of the applications listed, the switching
system will need to be able to access the frame structure of
the data, Fig.1, and operate at the high data rates carried by
the network. The switches themselves will need to operate with
fast switching times. This is an important difference between
the requirements of a communications network and the stated
requirements of an optical computing system. It is frequently

claimed that the low operating speed of optical logic elements can be compensated by a high level of parallel operation. This clearly cannot apply in communications applications where a high switching speed is essential for the operation of the system.

Apart from the constraints imposed by the network requirements, optical switching systems are also constrained by the specifications of existing optical fibre transmission systems. The wide bandwidths and low losses available with fibre optic transmission systems will not impose the limiting constraints on the network but the low optical power levels available within a fibre communication system may well limit some of the options available for switch design. If a 10^{-9} bit error rate is acceptable then a receiver sensitivity of 2000 photons/bit is required. For a 1 Gbit/s system this corresponds to -34dBm or 0.4 microwatts. These are power levels at the receiver so they are lower limits but we should not assume there to be more than 1-10 microwatts of optical power available for switching.

5.
OPTICAL SWITCHING DEVICES

We will consider five types of optical switch which have been proposed for the applications listed in section 3. These are:

(a) Circuit switches

These switches divert the light output from one fibre into one of N possible output ports. These are optical equivalents of the Strowger type switch used in the early automatic telephone exchanges. Several versions of these have been demonstrated involving electro-mechanical movement of fibres, switching with liquid crystal materials, beam deflection by a magnetic stripe domain element, and nonlinear optical optical switching using ZnSe filters. Descriptions of all these are given in reference [2]. These switches can be used in applications such as video

links, or reconfiguration of a faulty network but they do not
operate on a fast enough time scale to access formated data   in
the data stream. Fig.2   shows schematically the deflection
system developed at Sperry[3].

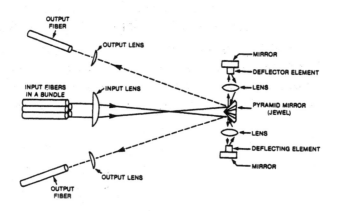

Fig. 2

(b) Four-port directional coupler switches and switch matrices

The basic structure of the directional coupler is shown in Fig
3a. Light launched into port A will couple to either D or E
depending on the optical length of the coupling region.
$Ti:LiNbO_3$ is the most common waveguiding material used for this
application. Typical characteristics are shown below:

Switching Speed > 1 GHz

Switching voltage - 10 volts

Cross talk > -20dB

Device length - 10 mm

Most recent efforts have been devoted to fabrication of arrays
of switches to produce cross point matrices such as that shown

schematically in Fig 3b.   These matrices are electrically
controlled and any input can communicate with any output using
a simple setting algoithm.   Although the characteristics listed
above are attractive, there are severe limitations on the
number of cross-points which can be operated in a single
matrix.   The limit arises because of constraints on the
substrate size and because of the difficulties of electrically
driving each cross point at the high frequency needed.
Typically 8 x 8 matrices are feasible and 16 x 16 matrices have
been demonstrated using a reflective design and reducing the
switch length to 3 mm (at the expense of a 25 -30 v drive) [4].

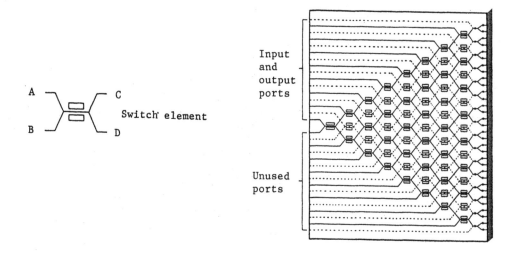

Fig 3

There are additional problems associated with the fact that
number of cross-points used will depend on the interconnection
specified. Since there is inevitably some optical loss at each
cross point, the insertion loss of the switching matrix will
vary depending on the interconnection specified. This is
undesirable when considering the total system design.   The
number of cross-points can be reduced if an alternative
interconnection scheme is used such as a perfect shuffle (see
Professor Lohmann's lectures in these proceedings). However the

setting algorithm to establish a particular interconnect is
more complex requiring a re-setting of all N^2 elements of the N
x N matrix. In view of these contraints it seems likely that
LiNbO$_3$ cross point arrays will find application only when a
small switching matrix is required.

(c) Active path optical switches with electrical control

The approach here is to split all input channels and route them
to all output channels. Laser amplifiers are placed in all
paths connecting input and output channels and amplification or
attenuation is introduced into the paths to be linked. The idea
is an attractive one but it suffers the disadvantage that the
amplifiers will have to be switched rapidly and will dissipate
quite a lot of power. The problem of driving large numbers of
these switches at high speed is daunting and, like the lithium
niobate switches, seem destined for applications involving
small arrays. Fig. 4 illustrates schematically the switching
configuration used [5].

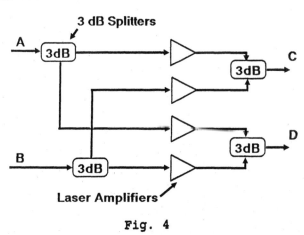

Fig. 4

(d) Optical logical devices for switching

So far the optical switching elements proposed have all had
electrical control. All optical switching is attractive in
concept and much effort has been dedicated to the development
of digital optical circuits which could be used in a digital

switching system. (See some of the earlier chapters in this book). Fig. 5 illustrates how an exchange-bypass circuit could be formed using logic gates and controlled optically via the C and C, if all inputs to the gates are assumed to be optical ones.

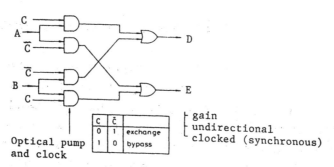

Fig. 5

The detailed characteristics of all-optical bistable devices and hybrid bistable devices have been listed in previous chapters in this book so we will not repeat them here. However we do note the following points:

- Devices currently available do not meet the specifications desired for applications listed previously, neither from speed considerations nor on the basis of power levels needed for switching.

- The hybrid SEED devices look the most promising devices and the performance demonstrated by the symmetric SEED, 1 nsec switching time and 1.5 mW switching power, indicates that if optical or opto-electronic logic does find an application in photonic switching , it is this type of device which will be used.

- Ultra-fast all optical devices may offer switching on a time scale of 100 fsecs. Electronic switching devices could not compete on this time scale but very high optical powers are required for these optical switches and considerable development is required.

(e) Optically controlled electronic switches

The search for suitable optical switches is driven by the fact that electronic switching systems do not yet have the speed capability to cope with the projected demands of networks such as the ISDN. The use of optically controlled elctronic switches is a rather different way of involving optics in the switching process. The approach is based on the idea that optics may be used to interconnect different regions on a wafer or interconnect small very high speed electonic modules. We believe that this hybrid approach offers many attractions particularly for wide-band switching matrices so we will discuss this in detail in the next section.

6.
A HYBRID APPROACH TO PHOTONIC SWITCHING

The design of very high speed electronic circuits is difficult, not because of individual device limitations, but because of difficulties in communicating information between devices some of which may be a short distance apart and some of which may be a considerable distance apart[6]. Problems of clock skew occur due to the fact that interconnection lengths may vary significantly. Closely spaced devices and transmission lines suffer from crosstalk which is difficult to predict in complex geometry layouts at high bit rates > 1 Gbit/sec .Other problems which arise when communication high data rate information electronically are associated with the dispersive properties, limited bandwidth and attenuation of metallic interconnections. These are problems which optics solves effectively. Optical communication can occur at high data rates without significant dispersion, loss or crosstalk and optical imaging can interconnect without introducing time skew. If the optical system involves free space interconnect then the possibility of parallel interconnection involving crossed optical paths may give additional advantages. It is important to note that this may just mean more flexible design using all three dimensions

for optical wiring, but it may also mean more powerful design
if the interconnection path itself performs some form of
processing. In the simplest case this may be just an inversion
or length scaling (e.g. magnification). In more complex
arrangements this may mean a shuffle of the data inputs from
one plane to another or a Fourier transform of inputs between
planes.

The essential feature of the hybrid approach is that
electronics is used to carry out the logical or decision
operations that it can perform well, while optics is used to
carry out the interconnection functions for which it is best
suited.   To implement such a scheme the electronic circuit is
split up into a number of modules. Each module, or "island "
contains a number of high speed electronic devices which are
interconnected only over short paths.   Interconnection between
modules is by optical means , Fig. 6, and can involve a spatial
reorganisation of the input data sources.   In any practical
implementation based on this approach we need to consider a
means of producing the required optical interconnection e.g.
bulk optics or holographic optics, and we need to select a
means of interfacing between optics and electronics e.g.
through detectors, sources or modulators which can be located
on-chip.

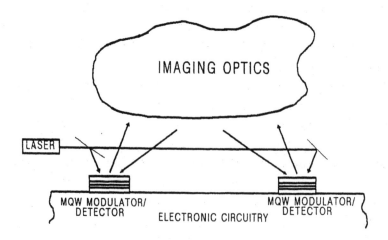

Fig. 6

7.

THE ELECTRONICS-OPTICS INTERFACE

The interface from optics to electronics is straightforward -
optical detectors perform this function and detector technology
is advanced both in silicon and III-V materials. The design of
on-chip receiver circuitry does require further consideration
and indeed may be the limiting factor in the complete design.
The interface from electronics to optics could involve lasers,
LEDs or modulators. For free space interconnect we would like
an interface element that transmits or reflects in a plane
perpendicular to the circuit plane, an element which can be
integrated with electronic components and one which can be
modulated at a high enough data rate to make the use of optical
interconnection worthwhile. The element must also operate at a
low bias voltage (certainly not more than 5 volts) so that is
can be driven from voltage rails normally present in electronic
integrated circuits. The element should consume little power.
The semiconductor laser does satisfy many of these requirements
but it difficult to arrange transmission out of the plane of
the circuit (although surface emitting laser technology is
advancing to meet this requirement). The main disadvantages of
semiconductor lasers is that the electronic to optical
conversion is relatively inefficinet so that power lost would
be dissipated on chip, resulting in large local variations in
temperature on the chip. This could cause many problems
particularly if the lasers were monolithically integrated on
silicon substrates. The development of III-V on silicon growth
techniques has been impressive and lasers have been
demonstrated by several groups but there are still lifetime
problems arising from the large difference in lattice constant
between GaAs and silicon and because of the mismatch in thermal
properties. It is this last feature which would be of most
concern when power is dissipated on-chip. The use of LEDs is
unlikely mainly because of the limited data rates possible, but
again thermal problems could also be significant.

Until recently the use of optical modulators could also have
been dismissed because of size or voltage considerations.

However the GaAs multiple quantum well electro-absorption
modulators described in D A B Miller's lecture appear to have
overcome many of the disadvantages of other electro-optic
modulators. Indeed a detailed study of this device, considered
as an interface for optical interconnection, shows it has most
of the features required of an interface element: small size,
high speed, low voltage, large electro-absorption change. If
used with a reflection stack and addressed by an off-axis
laser, the transmission of data naturally occurs perpendicular
to the circuit plane and most of the unwanted thermal energy is
dissipated well away from the chip.

There are two important factors which have to be
considered. Although the drive voltages are low and in
waveguides may be 1-2 volts, they are generally rather higher
than desirable when used in the reflection mode described.
Voltages quoted for 100 A wells are typically 10-12 volts. The
second factor which may be of concern is the fact that the
modulation depth (or contrast) is usually limited to about 3-4
dB. What are the possibilities of overcoming these difficulties?
One possibility for controlling the voltage is simply to
optimise the device performance subject to constraint e.g. a
maximum operating voltage of 5 volts. Using this procedure we
have found that it should be possible to obtain a transmission
change of 42% , a contrast of 3.3dB for a 5 volt drive
voltage[7]. Further improvements may be possible by a using
narrow well width for the quantum wells. We have reported
previously [8] that a wide well sample (145A) showed improved
performance at low voltages compared with narrow well samples
In the table shown below we list the performance achieved with
that device at very low voltages.

Drive voltage	Contrast	Transmission Change
2 v	1.5 dB	13%
4 v	1.9 dB	17%
6 v	2.3 dB	21%

This structure was not optimised so it can be anticipated that if the technique referred to above is used to optimse the device, further improvements may be possible. We can be confident that low drive voltages will possible provided that a low contrast is acceptable. The low contrast would be a problem if the device were acting as a switch but this is not the case here. The device is being considered as an interface element for an optically interconnected electronic switch and in this case a low modulation depth or contrast may simply introduce a power penalty. External modulation with a low extinction ratio lithium niobate modulator has shown this to be the case [9].

8.
A SELF ROUTEING WIDEBAND SWITCHING MATRIX

We will consider how the hybrid design can be used to produce a wideband switching matrix suitable for routeing high speed packet data through a switch. The architecture is similar to that used in the Starlite[10] switch but in this case is discussed in terms of optical interconnection. The task to be carried out can be stated as follows. Data enters the switch on a one dimensional array of optical fibres. Each data stream carries an address which indicates the output port to which the data must be routed - irrespective of the data port at which the data arrives. The output fibres can then collect the output data sorted for distribution to the correct destination. The optical scheme which we present here has been described in detail elsewhere [11], so we will concentrate on the principles rather than the details.

The algorthm proposed for the sorting operation is the perfect shuffle sort algorithm [12]. The algorithm will sort a one dimensional array of input addresses into an ordered form - the highest address appearing at one end of the array and the lowest at the other end. The algorithm involves a sequence of logiacal operations followed by a shuffle interconnect followed by another logical operation and another shuffle, and so forth. The logical operations are exchange or bypass operations (the two inputs from neighbouring data inputs are compared and, depending on the relative magnitudes, may be re-ordered (exchanged) or left unchanged (bypass). The shuffle interconnection involves taking the spatial input data, dividing it in two, magnifying the spatial separation by two and overlaying. That is, if we label a pixel element by j in the one dimensional array of N spatially separated inputs, then j will map to $k = 2j - 1$ for j in the range 1 to N/2 and $k = N - 2(N - j)$ for j in the range N/2 + 1 to N. The attraction of the perfect shuffle sort algorithm is that the shuffle operation can be implemented optically, as discussed in Professor Lohmann's lecture. The logical operations can be carried out electronically with the quantum well modulators acting as interface elements. The sequence of operations is illustrated in the diagram below.

INPUT DATA (numbers indicates address of output)	4	2	1	3
1st logical operations		+		−
	2	4	3	1
1st shuffle				
	2	3	4	1
2nd logical operations		+		+
	2	3	1	4
2nd shuffle				
	2	1	3	4
3rd logical operations		+		+
OUTPUT DATA	1	2	3	4

The detailed analysis of this switch [11] for 128 input ports, shows that a matrix of exchange - bypass modules 43 x 64 would be required, involving 42 shuffle operations. It is suggested that a data rate of 1 Gbit/s would be possible and would require approximately 300 microwatts of optical power per modulator. The total optical optical power of 860 milliwatts could be provided by an array of semiconductor lasers.

9.

CONCLUSIONS

At the present time it is difficult to see how optics can become the dominating technology for switching systems. There are some clear niches in which optics will have a role: circuit switches, small dimensioned cross-point switches in lithium niobate for example. However electronic switching would appear to offer equal performance for many applications. We suggest that the hybrid approach, in which optics assists electronics, is the most likely way for optics to extend the performance of switching systems.

REFERENCES

1. J.E. Midwinter, J. Lightwave Tech, 6, 10 (1988).

2. Technical Digest of "Topical Meeting on Photonic Switching", IEEE/Optical Society of America, March 1987.

3. E.J. Torok, J.A. Krawcsak, G.L. Nelson, B.S. Fritz, W.A. Harvey and F.G. Hewitt, Technical Digest of "Topical Meeting of Photonic Switching, IEEE/Optical Society of America, p49, March 1987.

4. P.J. Duthie, M.J. Wale, I. Bennion, Technical Digest of "Topical Meeting on Photonic Switching, IEEE/Optical Society of America, p71, March 1987.

5. R.M. Jopson and G. Eisentein, IEEE/OSA Topical Meeting on Photonic Switching, Incline Village, NV, March 1987.

6. J.E. Midwinter, Physics in Technology, 19, pp101-8 and 153-165, 1988.

7. P.J. Stevens and G. Parry, to be published in J. Lightwave Tech. 1989.

8. M. Whitehead, P. Stevens, A. Rivers, G. Parry, J.S. Roberts, P. Mistry, M. Pate and G. Hills, Appl. Phys. Lett., 53, 956, 1988.

9. J-M.P. Delavaux, C.Y. Kuo, T.V. Nguyen, R.W. Smith, Elect. Lett., 22 pp1139-1141, 1986.

10. A. Huang and S. Knaner, Proc. IEEE Global Telecommunications Conf, Atlanta, Georgia, USA, (New York IEEE) pp121-125, 1984.

11. J.E. Midwinter, Proc. Inst. Elec. Eng., 134 pt. J, pp261-268, 1987.

12. H.S. Stone, IEEE Trans., C-20 pp153-161, 1971

PRESENT COMPUTER ARCHITECTURES AND THE POTENTIAL OF OPTICS

Karl-Heinz Brenner

Applied Optics, University of Erlangen

1.

HISTORY

1930 Ch. Babbage (UK) develloped a concept for a mechanical programmable computer.

1936 K. Zuse (GER) built a mechanical programmable computer at his home.

1937 Aiken proposed the "Automatic Sequence Controlled Calculator, Mark 1". It was built 1939-44 at IBM.

WW2 several electromechanical and electric computers for warfare (USA).

1946 Burks, Goldstine, von Neumann: First concept for a general purpo;e computer. A breakthrough not by technology but by theoretical works.

1948 Eckert, Mauchly build ENIAC, an electronic version (tubes) of Aikens' Mark 1.

1949 Wilkes (UK) build an electronic computer based on J. von
     Neumann's concept.

1951 von Neumann's computer (EDVAC) reached completion. He
     started design of IAS.

1951 Eckert and Mauchly build UNIVAC : the first commercial
     computer.

2.

THE CONCEPT OF BURKS, GOLDSTINE AND VON NEUMANN[1]

1.   The computer is divided into four logically and
     spatially separate units:

     a: an arithmetic unit for arithmetic and logic
     functions.

     b: a memory, where programs and data are stored.

     c: a control unit

     d: an input/output unit where data and programs are
     entered and retrieved.

2.   The computer in its structure is independant of the
     problems to be solved. Only the program, coming from
     outside the machine determines the operation.

3.   Programs and data reside in the same memory.
     (Stored program computers)

4.   The memory consists of individually addressable words.

5.   Instructions are mostly stored in subsequent memory
     locations

6.   There are jump-instructions

7.   There are conditional jump-instructions

8.   The binary number system is used.

Today, these concepts sound so evident because all the
existing single processor computers (PDP, VAX, u-Processors,
etc.) are built based on these concepts. These concepts were
not at all evident at that time.

3.

THE STRUCTURE OF VON NEUMANN'S COMPUTER

The first von Neumann computer [2,3] was build with a memory
of 2^{12} = 4096 words each consisting of 40 bits. von Neumann
himself intended to use a cathode ray storage tube as memory.
He envisaged an access time of 5 - 50 us. The actual memory
that has been uced then was a magnetic core memory.

    The arithmetic unit was able to perform adds and shifts
and contained two registers A and B, a program counter P and
an instruction register I. A condition sensing logic C
enabled conditional jumps. The structure of an instruction
was:

    OPERATION address

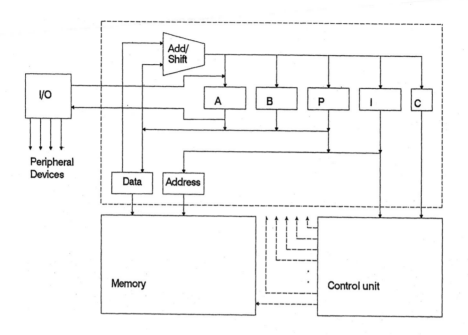

Fig. 1 The general structure of a von Neumann computer

4. Further developments

Arithmetic & Logic Unit (ALU):

    multiplier hardware.
    fast adders.
    floating point arithmetic.

    Many ALU's work in parallel
    (array computers, ILLIAC IV: 64 ALU's)

    One   ALU   is   time-shared   among   different   parallel
    processes

    Pipelining (temporal parallelism)

Control unit (CU):

indexing: the address is not fixed but modifiable by an offset in a register. Thus addressing an array is possible.

indirection: the address is not part of the instruction but is read from a memory cell. (Effective address calculation)

reusing instructions: In the inital concept reusing instructions was only possible by two mechanisms: the loop (consisting of a sequence of instructions and a conditional jump) and by the mechanism of self modifying code. The subroutine concept together with the stack has been added later. It provided modularity in the programs.

micro programming: This concept allows to change the instruction set in the same way as changing the program. Thus instructions adapted to special problems could be designed.

Input/Output:
serial, parallel, Direct Memory Access

Memory:
1950's   : rotating magnetic drums
late 50's : magnetic core memory
70's   : integrated circuit memory (static, dynamic)

memory hierarchy (virtual memory)

independent programs in different segments of the memory (multi programming, address translation)

5.

PARALLEL COMPUTERS

Parallelism can be introduced replicating some of the units
in a computer. For the case of replicating the arithmetic
logic units, a classification was given by Feng[4]:

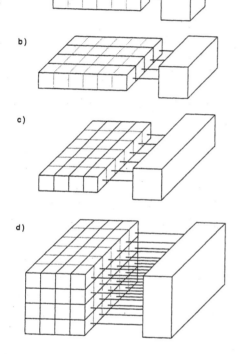

a)

b)

c)

d)

Fig. 4

words are processed in
series and bit serial.
Word serial/ Bit serial

words are processed in
parallel but bit serial.
Word serial/ Bit parallel

words are processed in
series but bit parallel.
Word parallel/Bitparallel

words and bits are pro-
cessed in parallel.

For parallelism by replicating both the ALU and the CU
a classification scheme is given by Flynn[5]:

Instructions \ Data	SINGLE	MULTIPLE
SINGLE	SISD	SIMD
MULTIPLE	MISD	MIMD

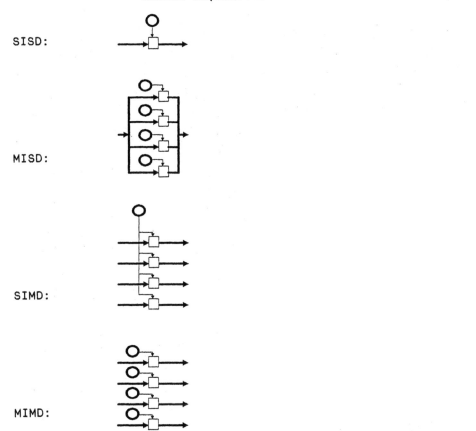

SISD:

MISD:

SIMD:

MIMD:

Fig. 3 Processors (box) are supplied with data (thick arrows) and instructions (circle).

The Erlangen classification system (ECS) by Händler[6] expresses the pipelineing and parallelism occuring on all levels of a computer system:

$$T(C) = \langle\ K \times K',\quad D \times D',\quad W \times W'\ \rangle$$

K : # of control units
K': # of control units that can be pipelined
D : # of ALU's controlled by each CU
D': # of ALU's that can be pipelined
W : # of bits in an ALU (wordlength)
W': # of pipeline segments in all ALUs or in a single PE

In a pipeline, a task is decomposed into several elementary simple tasks. Each simple task is assigned to a segment of the pipeline. The processing time for one data packet is determined by the latency (# of segments) of the pipeline. The throughput as the processing rate is however increased by the number of segments. This can be considered a temporal rather than a spatial parallelism. Computers of this kind are called vector processors whereas processors with spatial parallelism are called array processors.

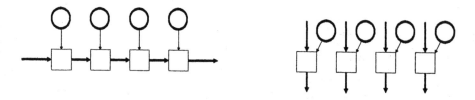

Fig. 4 Processing in a vector processor and in an array processor.

Examples :

   SIMD: <1, N, 1> (optically <1, N², 1>)

   Cray-I: <1, 12x8, 64x(1-14)>

The Cray-I is a single CPU computer with 12 pipelined functional units. Up to 8 functional units can be chained together to form a pipeline. The functional units have from 1 - 14 segments.

What is still missing in the existing classification schemes:

      - Connectivity of data streams (data driven, systolic)

      - Connectivity of processors

Connection schemes:

chain

ring

star

tree

mesh, torus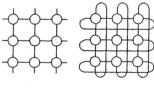

permutation network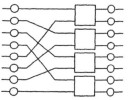

N-dim. Hypercube

fully interconnected network

Present status of electronic computers:

- Supercomputers : ≈ 100 MHz

- Connection machine : 64000 Processors

6.
THE POTENTIAL OF OPTICS

6.1 The 10^6-pin connector

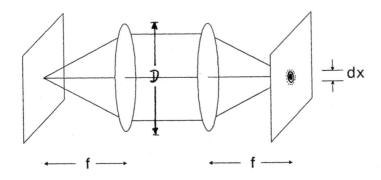

Fig. 5 An optical imaging system

   if $O(x) = \delta(x)$; Object

      $I(x) = |\ 1/\pi x\ \sin(\pi x D/\lambda f)\ |^2$; Image

The width of one resolvable spot is : $dx = 2\lambda f/D$
If imaging with low distortion is possible in a range

      $-D/4 \leq x \leq D/4$ ⇒ $\Delta x = D/2$

The number of resolvable spots is

$$N = \Delta x/dx = D^2/4\lambda f$$

With D = 3 cm, f = 16 cm, λ = 600 nm $\Rightarrow$ N = 2300x2300 pixels assuming diffraction limited optics and a space invariant (SIV) interconnection. In the case of space variant interconnections (SV) the number of resolvable elements is given by

$$N_{SV} = \sqrt{N_{SIV}}$$

Interconnections can be categorized not only with respect to space variance but also with respect to fan-out:

space invariant (regular)   space variant (irregular)

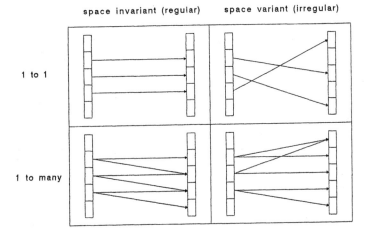

Fig. 6 Different types of interconnect structures

6.2. > 10^9 - Hertz Operation

Limitation due to propagation (Dispersion)

$$f < 1/8\pi \sqrt{\frac{c}{D\,L\,N}} \quad ; \quad D = \frac{dn}{d\omega} \quad \text{Dispersion}$$

N: # of harmonics to be propagated
L: propagation length

in glass: > 100 GHz
in air: ≈ 100 THz

Limitations due to devices:
  depend on
  - material
  - device design (size, cooling)
  - device operation (critical slowing down, ..)

6.3. Optical isolation
  Photons do not interact in linear media -

6.4 Channel capacity

   Currently the data rate is only limited by the devices.
1 GHz operation seems feasable. Combining temporal and
spatial bandwidth, one could define the optical channel
capacity:

  $C = N f$ (in bits /sec)
  $C = 10^6 \ 10^9 = 10^{15}$ bits / sec

7.

POWER CONSIDERATIONS

According to Feynmann, computing itself can be done without
energy. If speed is of any concern (practical situation)
computing does require energy.

The elementary quantity is the energy per area, required to
do one switching operation, E_s. Lowest values today are:

  $E_s = 10$ fJ/(um)^2

The power per area is the product of the frequency f and the energy E_s. For an operation in the thermal equilibrium, this power has to be dissipated. Typical values for dissipating thermal power are currently

$$P_D = 10^{-7} \text{ W}/(\text{um})^2$$

resulting in a maximum operating frequency of

$$f = P/E_s = 10^7 \text{ Hz}$$

if the devices were packed densely. To increase the device speed one has to space the devices further apart. With a packing factor p defined as the ratio of total device area to active area the device speed can be increased by a factor of p. With the available space bandwidth product limited (typically 10^6), the number of devices will however decrease by a factor of p. Thus the channel capacity is constant with bulk lens imaging systems. Possible tradeoffs are 10^6 devices at 10^7 Hertz or 10^4 devices at 10^9 Hertz.

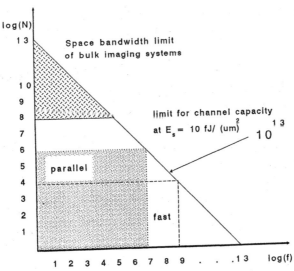

Fig. 7 illustrates the tradeoff between superfast and massively parallel operation.

The constant channel capacity can be increased by either increasing the available space bandwidth product, by decreasing the device energy or by increasing the cooling rate. The cooling rate could possibly be increased by an order of magnitude. The device energy cannot be decreased much further because this value is already close to the statistical limit. The space bandwidth product cannot be increased with bulk lenses but it could with lenslet arrays because there the area limitation is removed.

References:

1. A.W. Burks, H.H. Goldstine, J.v.Neumann: US Army Ordnance Department Report, 1946. printed also in: C.G. Bell and A. Newell, "Computer structures: readings and examples", New York Mc Graw Hill 1971, p. 92-119.

2. F.J. Hill, G.R. Peterson: "Digital Systems: Hardware and Design", John Wiley & Sons, New York 1973

3. W. Händler: "Digitale Universalrechenautomaten", Aus Taschenbuch der Nachrichtenverarbeitung, K. Steinbuch ed. Springer Berlin-Heidelberg-NewYork 1967

4. T.Y. Feng: "Parallel computers and processing", ACM Comp. surveys, 9 (1977)

5. M.J. Flynn: "Some computer organisation and their effectiveness" IEEE Trans. on comp. C21, 948 (1972)

6. W. Händler: "On classification schemes for computer systems in the post-von Neumann era", in Lecture Notes in Computer Science, 26, October 1974, Springer Verlag

OPTICAL LOGIC AND ARITHMETIK

Karl-Heinz Brenner

Applied Optics, Univerity of Erlangen, FRG

1.
NUMBER REPRESENTATIONS

In the oldest number systems, numbers were represented by enumeration. For large numbers, this representaion is not optimal. The romans improved the system by grouping:

```
IIIII := V    VV    := X
XXXXX := L    LL    := C
CCCCC := D    DD    := M
```

Our modern number system, the g-adic syotem, originated in india and came to us through the orient (Arabian numbers). It can be represented as a power series:

$$n = \sum c_n \, g^n \tag{1}$$

In our system, g=10 requiring an alphabet of ten symbols:

$$c_n \in \{ 0, 1, 2, 3, 4, 5, 6, 7, 8, 9 \} \tag{2}$$

The binary number system:

With the base g = 2, only two different symbols are necessary:

$$n = \sum c_n \, 2^n \; ; \quad c_n \in \{ 0, 1 \} \tag{3}$$

making it attractive for computers. The hexadecimal system, which is used to represent binary numbers in a shorter form has a base g=16 and 16 symbols:

$$n = \sum c_n \, 16^n; \quad c_n \in \{0,1,2,3,4,5,6,7,8,9,A,B,C,D,E,F\} \tag{4}$$

Residue numbers:

It can be shown that a finite set of numbers can be represented uniquely by an k-tupel of remainders:

$$n = \{m_0, m_1, \ldots m_k\} \text{ with } m_j = n \bmod p_j \, , \; j=0..k \tag{5}$$

if the moduli p_k are prime to each other. Under these circumstances numbers can be represented uniquely in the range:

$$0 \leq n < \prod_{j=0}^{k} p_j \tag{6}$$

The advantage of this representation is that the speed of the arithmetic operations $\oplus = \{ +, \; -, \; * \}$ can be independant of the size of the numbers because these operations can be performed in parallel:

$$\tag{7}$$
$$\{x_0, x_1, \ldots x_k\} \oplus \{y_0, y_1, \ldots y_k\} = \{x_0 \oplus y_0, x_1 \oplus y_1, \ldots x_k \oplus y_k\}$$

It always takes one step to perform the operation. No "carry" has
to be considered as in g-adic systems:

The disadvantage of this number system is that there are no
simple algorithms for:

- conversion to decimal (binary)
- division
- comparison

Modified signed digit number representation[1]:

This representation offers also fully parallel, carry-free
operation. Numbers are represented similar to the binary
representation in a polynomial form but with the three
symbols {0, 1, $\bar{1}$}. 0 and 1 have the usual meaning and $\bar{1}$
has the meaning of -1. As a result this representation is no
longer unique, i.e. one number can have different representations. The addition is performed carry-free in three
steps. In the first step, a carry is allowed, however the
following additional rule for addition is applied:

$$1 \times 2^n + 0 \times 2^n = 1 \times 2^{n+1} + \bar{1} \times 2^n \qquad (8)$$

In step two and step three the regular addition rules are
applied. This guarantees that no carry can propagate further
than one position. As a result an addition always takes
exactly three steps, independent of the word length.

The general disadvantage of multi-valued versus two-valued

systems is that multi-valued systems need larger logic tables
that have to be realized by hardware.

Example of addition:

```
        0  1  2  3  4  .  .

  0     0  1  2  3  4
  1     1  2  3  4  5
  2     2  3  4  5  6
  3     3  4  5  6  7
  .
  .
```

2.
OPTICAL CODING OF INFORMATION

Intensity coding:

With intensity coding, a value v is determined by comparisons
to equidistant thresholds:

$$v = \begin{cases} 0 & : \quad\quad\quad I < T_0 \\ 1 & : \quad T_0 \leq I < T_1 \\ . \\ . \\ n-1 & : \quad T_{n-2} \leq I < T_{n-1} \end{cases} \quad\quad (9)$$

In the special case of n=2 there is only one threshold
determining if the number is 0 or 1.

Positional coding:

In positional coding, the position of a light spot determines
the value of a number. If x is the position of the light

spot, then

$$
v = \begin{cases} 0 & : x = x_0 \\ 1 & : x = x_1 \\ . & \\ . & \\ n-1 & : x = x_{n-1} \end{cases} \tag{10}
$$

The special case n=2 is called dual rail coding, because the
'0'-pattern is the complement of the '1'-pattern:

Polarization coding:

Using the polarization of light, 4 states can be
distinguished assuming a constant field amplitude $E_x = E_y$:

$$
\begin{pmatrix} E_x \\ 0 \end{pmatrix} \triangleq \ '0' \qquad \begin{pmatrix} 0 \\ E_y \end{pmatrix} \triangleq \ '1' \qquad \begin{pmatrix} 0 \\ 0 \end{pmatrix} \triangleq \ 'A' \qquad \begin{pmatrix} E_x \\ E_y \end{pmatrix} \triangleq \ 'B'
$$

Phase coding:

With phase coding, the relative phase with respect to a
reference phase would be used for coding different states. It
is very sensitive to thermal mechanical influences.

Wavelength coding:

Different wavelength can be used for coding different states.
The necessary linewidth is related to the bandwidth

$$
\Delta\lambda = \lambda/f \ \Delta f \ ; \quad f : \text{frequency}, \ \lambda : \text{wavelength} \tag{11}
$$

This coding however requires an interaction mechanism between different wavelength. So, for a NAND-function for example, a gate with λ_1 at both inputs has to produce λ_0 at the output.

3.
OPTICAL IMPLEMENTATION OF LOGIC

Logic is the basis for realizing numeric or nonnumeric operations. Implementation of logic, in this definition includes the hardware aspects (devices and interconnects) and the underlying logic:

Logic Implementation = (Logic, Device, Interconnect)

Existing types of logic:

Boolean logic (BL):

is based on the set of symbols {0, 1} and operators {∧,∨}. The satisfy the relations:

$$a \wedge 0 = 0, \quad a \vee 0 = a$$
$$a \wedge 1 = a, \quad a \vee 1 = 1 \qquad a \in \{ 0, 1 \} \qquad (12)$$

and DeMorgan's Law:

$$\overline{a \wedge b} = \bar{a} \vee \bar{b}, \quad \overline{a \vee b} = \bar{a} \wedge \bar{b} \qquad (13)$$

Binary image algebra (BIA)[2]:

originated from Mathematical Morphology[3]. It is defined by the set of elementary images { a, ā, b, b̄ } and the set of operators {⁻, ∪, ⊕ }. The elementary images are the four positions (0,1), (0,-1), (1,0) and (-1,0).

The operators are:

⁻ : the complement operator $\bar{X}$ = { x ∈ Z², x ∉ X }

∪ : set union   X∪Y = { x : x ∈ X ∧ x ∈ Y }

⊕ : image dilation   X⊕Y = { x+y : x ∈ X, y ∈ Y}

It is clear that Binary Image Algebra is especially adapted to parallel operations on two-dimensional images.

Cellular Logic (CL):

like BIA is defined on higher dimensional spaces. It goes back to J. von Neumann, who used it to construct self replicating automata. CL can be described by a neighborhood and a transition function. According to Toffoli[4] it is defined as a structure:

$$CL = (A, S, X, \lambda), \text{ where} \qquad\qquad (14)$$

A is a set, the state alphabet (typically {0, 1}

S is a free abelian Tesselation group (the space)

X the neighborhood template

$\lambda : A^X \to A$

is a mapping of the state of the neighbors to the state alphabet

Typical neighborhoods are the von Neumann neighborhood (center, north, south, east, west} or the Moore neighborhood that also includes diagonal neighbors. The characteristic is that the state of <u>one</u> cell is determined by its own previous state and the state of the neighborhood.

Symbolic Substitution (SS):

has been suggested by A. Huang[5,6] as way to utilize the
parallelism of optics. It can be defined as a structure:

SS = (A, S, N, φ, Ω), where   (15)

A is a state alphabet ({ 0, 1 })
S is a space (typically Z^2 or a subspace of it)
N is a pair (L $\in$ S[1], R $\in$ S[r]) of neighborhood patterns
φ : A[1]x S[1] $\to$ A[r]x S[r]
is a local mapping that maps a left-hand side pattern onto
a right-hand side pattern.

Ω is a superposition operation for the global mapping

The essential difference of SS to CL is that φ is not a
mapping into the state alphabet but rather a mapping onto
a pattern and that SS involves a superposition operation.

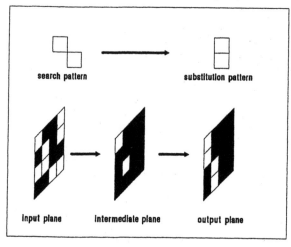

Fig. 1 Principle of Symbolic substitution

We want to note that each logic is complete. If they weren't,
they would not be very useful. As a result of the complete-

ness, every logic can be expressed by any other logic. So for example Boolean logic can be implemented by Symbolic substitution or Symbolic substitution can be expressed by Binary image algebra and so on. None is a superset of the others. Which logic is more useful depends on the application.

List of existing devices:

    Liqid crystal light valve (LCLV) (FLC)
    Acousto-optic light modulator (AOM)
    Waveguide modulators (WM)
    Optoelectronic devices (OED)
    Nonlinear Fabry Perots (NLFP)
    Self Electrooptic Effect Devices (SEED, D-, T-, S-, ..)

Interconnects:

    Optical fibres (OF)
    Waveguides (WG)
    Holographic interconnects (HOI)
    Classical optical interconnects (COI)

Current implementations of optical logic can now be expressed by the three-tupel (L, D, I). The cube in the following figure includes all possible and impossibles kinds of optical logic that can be constructed by combination.

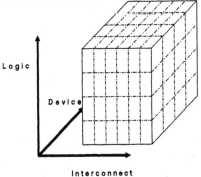

Fig. 2 Cube of possible logic implementations

Acoustooptical multipliers: LI = (BL, AOM, COI)

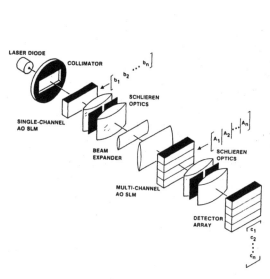

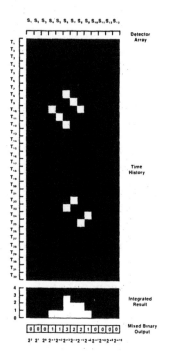

Floating-point acousto-optical matrix-vector multiplier.

Inner product mixed binary output.

R. Bocker[7] et al. 1985

Binary multiplication is performed by a convolution of the bit patterns. The result appears in the so called 'mixed binary' form that has to be converted to binary form involving a carry propagation.

Sequential optical logic[8]: LI = (BL, LCLV, HOI)

In this approach a logic circuit is converted into a form that uses only NOR-gates. The LCLV is used as the NOR-gate. Holographical optical interconnects provide the space variant interconnection of the gates.

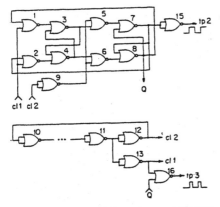

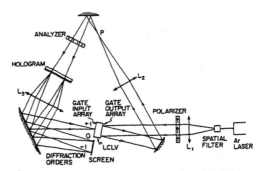

Fig. 23. Experimental system. Lens L_2 images from the LCLV gate output plane (LC plane) to the hologram plane. Lens L_3 provides a Fourier transform from the hologram plane to the LCLV gate input plane. The hologram comprises an array of subholograms.

Fig. 24. Test circuit consisting of a synchronous master–slave flip-flop with driving clock.

B.K. Jenkins[8] et al.

Hybrid combinatorial logic[9]: LI = (BL, AOM+ECL, COI)

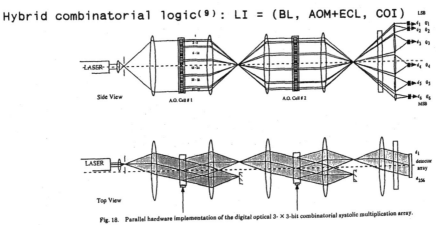

Fig. 18. Parallel hardware implementation of the digital optical 3- × 3-bit combinatorial systolic multiplication array.

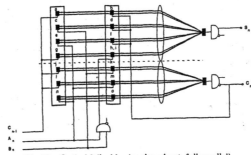

Fig. 10. Optical full adder (nonbroadcast, full parallel).

P. Guilfoyle et al.[9], 1988

Boolean equautions in this logic also are converted into a
form that only involves NOR-operations. The OR-operation is
done by optical superposition, the negation by electronic
hardware.

Restoring optical logic[10]: LI = (BL, NLFP, COI+HOI)

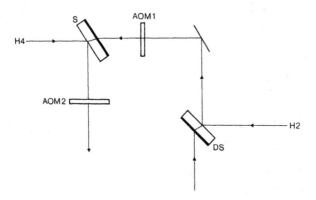

Fig. 7. Schematic of the optical configuration used to demonstrate that either of the adder outputs could be stored temporarily and then used as an input to a subsequent similar full-addition.
AOM-1,2 are acousto-optic modulators, S is a temporary store.

F.A.P. Tooley, et al.[10]

The NLFPs here are nonlinear ZnSe interference filters. Their
characteristic nonlinearity has been used to provide the sum-
and carry-operation necessary for an addition. The timing and
control using the acousto-optic modulators is achieved by a
mechanism called "lock and clock". Also other digital
circiuts like a ring oscillator and a multiplexer have been
demonstrated.

Logic by spatial filtering [11]: LI = (BL, LCLV, COI)

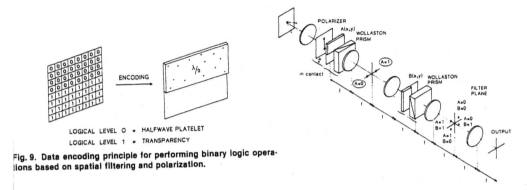

LOGICAL LEVEL O = HALFWAVE PLATELET
LOGICAL LEVEL 1 = TRANSPARENCY

Fig. 9. Data encoding principle for performing binary logic operations based on spatial filtering and polarization.

Fig. 10. 8-f spatial filtering setup for "polarization logic." The separation of the logical levels in the filter planes is achieved by Wollaston prisms.

A.W. Lohmann, J. Weigelt[11] 1987

Here the logic operation is not performed by the device but rather by a filtering operation. The device is used to code the input by polarisation. The logic states in the first plane will be separated horizontally by the first Wollaston prism. Thus in plane B the information about plane A is coded by direction. The four logic combinations of A and B are separated through the second Wollaston prism and in the subsequent Fourier plane all the possible logic combinations can be selected by filtering.

Shadow casting logic (12):

LI = (BL, mask or BSO, COI)

Like in spatial filtering logic, the logic device is used to code the input. Here the states in plane A are coded dual rail horizontally by a mask and in plane B dual rail vertically by a second mask. By off-axis illumination shifted copies of the superposed masks appear. The logic result in the centre of the 3x3 block can be selected to be any of the 16 possible Boolean functions with two inputs and one output by choosing the proper combination of illuminating LEDs.

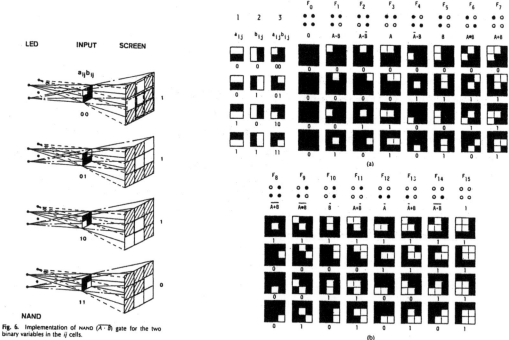

Fig. 6. Implementation of NAND ($\overline{A \cdot B}$) gate for the two binary variables in the ij cells.

Fig. 5. Optical representation of the 16 possible logic functions of the two binary variables with bright-true-logic. Columns F_0 through F_{15}, corresponding to the function names in Table 1, show the projections of code patterns in column 3 expressing the combinatorial variables with radiating configurations of the LEDs shown at the top. Bright-true-logic is indicated by combination of bright (1) and dark (0) in the central parts of the projections.

Tanida, Ichioka (12) 1984

Programmable logic arrays (PLA)[13]: LI = (BL, SEED, COI)

PLAs present a technique for designing and implementing digital logic circuits with a regular array of optical logic gates and regular optical interconnects. Because the structure is regular, any Boolean circiut can be customized by applying suitable masks. The interconnects are made with space-invariant free space optical components like beam splitters and lenses. The cost for regular interconnects is generally a larger gate count than with a circuit built with irregular interconnects, here however the benefits of regularity excede these costs.

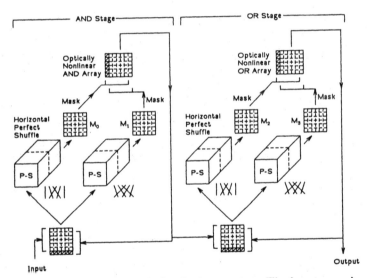

Fig. 1.   Schematic of a digital optical computer. The input array is split into two identical images which are each perfect shuffled (marked as *P-S*) and recombined onto an optically nonlinear array of AND gates. The AND array is imaged onto a similar setup with an OR array. Feedback is at two places as shown, and masks customize the interconnects so that only selected sites will be enabled.

M. Murdocca et al. [13] 1988

Cellular logic (14),

LI = (CL, LCLV, HOI)

Cellular logic requires the state of a cell to be a function
of its own state and the state of the neighborhood. A
hologram is used to bring together the neighborhood at the
center cell.

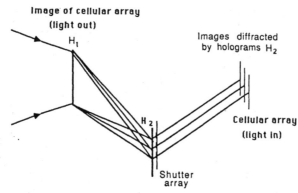

Fig. 6.   Two-hologram setup for programmable interconnect patterns. Only the first diffracted order of each hologram is shown.

J. Taboury et al. (14) 1988

Symbolic substitution (5,6,15): LI = (SS, OED or SEED, COI)

The mechanism of pattern recognition and substitution (Rec-
Sub) can be implemented with regular arrays of logic gates
and regular optical interconnects. It can serve to realize
Boolean circuits. It has been shown to perform array arithmetic in the binary number system, the modified signed digit
number system and in the reside number system. It has also
been shown to be capable of implementing Turing machines and
programmable processors. In the figure, a Rec-Sub operation
was demonstrated with a detector/LED combination as device.

K.-H. Brenner[15] 1987

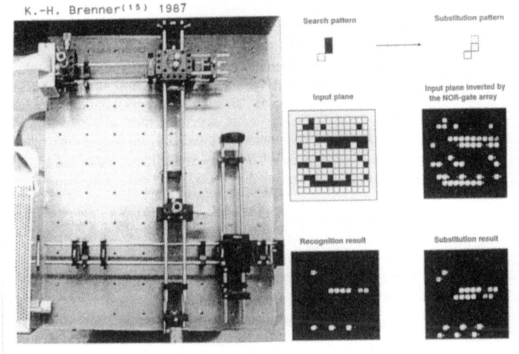

References:

1. R.P. Bocker, et al. :"Modified signed digit addition and
    subtraction using optical symbolic substitution", Appl.
    Optics 25, 2456 (1986)

2. R.M. Haralick et al. "Image analysis using Mathematical
    Morphology", IEEE Trans. on Pattern Analysis and Mach.
    Intell. PAMI-9, 532 (1987)

3. K.S. Huang et al.: "An image algebra representation of
    parallel optical binary arithmetic", submitted to Appl.
    Optics

4. T. Toffoli: "Cellular automata mechanics", University of
    Michigan, Techn. report No. 208 Nov. 1977

5. A. Huang : "Parallel Algorithms for optical digital

computers", Proc, of the 10th Intern. Optical Computing
Conf. 1983, IEEE Catalog No.: 83 CH 1880-4, p. 13

6. K.-H. Brenner, A. Huang, N. Streibl: "Digital Optical
    Computing with Symbolic Substitution", Appl. Optics, 25,
    3054 (1986)

7. R.P. Bocker, W.J. Miceli: "Optical matrix-vector multi-
    plication using floating point arithmetic", Top. Meeting
    on Opt. Computing, Techn. Digest Winter '85, OSA, TuD3-1

8. P.S. Guilfoyle, W.J. Wiley: "Combinatorial logic based
    digital optical computing", Appl. Optics 27, 1661 (1988)

9. B.K. Jenkins, A.A. Sawchuk et al.: "Sequential optical
    logic implementation", Appl. Optics 23, 3455 (1984)

10. F.A.P. Tooley et al.: "Experimental realisation of an
    all-optical single gate full adder" Opt. Comm. 63, 365
    (1987)

11. A.W Lohmann, J. Weigelt: "Spatial filtering Logic based
    on polarisation", Appl. Opt. 26, 131 (1987)

12. Y. Ichioka, J. Tanida: "Optical parallel logic gates
    using a shadow-casting system for optical digital
    computing", Proc. IEEE 72, 787 (1984)

13. M.J. Murdocca et. al.: "Optical design of programmable
    logic arrays", Appl. Optics 27, 1651 (1988)

14. J. Taboury et al.: "Optical cellular processor
    architecture. 1: principles", Appl. Opt. 27, 1643 (1988)

15. K.-H. Brenner: "Digital optical computing", Appl. Phys. B
    46, 111 (1988)

PROGRAMMABLE DIGITAL OPTICAL COMPUTERS

Karl-Heinz Brenner

Applied Optics, University of Erlangen, FRG

1.
FUNCTIONS AND SWITCHING FUNCTIONS

The characteristic property of a function is that the output at a certain time $t+\Delta t$ depends only on the state of the input at time t. If S(t) is the input (stimulus) and R(t) is the output (response) then

$$R(t+\Delta t) = F(S(t))$$

A function has no time spread (memory). Optical functions are functions of a two-dimensional input distribution S, typically a two-dimensional output distribution and mostly linear and space invariant. The most general function would be a mapping of an m-dimensional quantity onto an n-dimensional quantity which is nonlinear and space variant.

1.1 Boolean functions

If the lateral dimension is quantized and the amplitude is binary, we can speak of a boolean function. It can be

implemented by a circuit with m input channels and n output
channels.

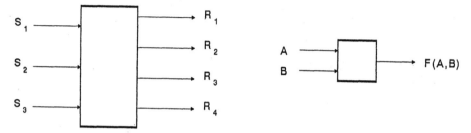

Fig. 1 Boolean functions

The special case of a logic gate (m=2, n=1) is shown at the
right hand side. The comlexity of a Boolean function is
finite and is determined by the number of possible functions:

$$C = 2^{n \times 2^m}$$

1.2 Description of a Boolean function

A Boolean function can be described by a normal form
enumeration all the cases where the response R_k is '1':

$$R_k = \bigvee_{R_k = 1} m_v$$

It is an OR-combination of all min-terms

$$m_v = (d_1\ S_1) \wedge (d_2\ S_2) \wedge \ ...\ (d_m\ S_m)$$

$$d_k\ S_k = \begin{cases} S_k & \text{if } d_k = 0 \\ \bar{S}_k & \text{if } d_k = 1 \end{cases}$$

A Boolean function can be realized by circuits or by a
lookup-table.

2.
STATE MACHINES

State machines[1] do not show the limitation of a Boolean function because of an additional internal state $Q(t)$. With the input being $S(t)$ and the output being $R(t)$ the state machine is fully described by the two equations:

$$R(t+\Delta t) = F \, (Q(t), \, S(t))$$
$$Q(t+\Delta t) = G \, (Q(t), \, S(t))$$

F and G are the defining functions of the state machine. If $Q(t)$ is not a continuous quantity but a finite set

$$Q = \{ \, q_1, \, q_2, \, \ldots \, q_1 \, \}$$

one speakes of a <u>finite state machine</u>. In the discrete case also S and R are sets:

$$S = \{ \, s_1, \, s_2, \, \ldots \, s_m \, \}$$
$$R = \{ \, r_1, \, r_2, \, \ldots \, r_n \, \}$$

and

$$F \, : \, S \times Q \to R$$
$$G \, : \, S \times Q \to Q$$

are mappings. A finite state machine can be specified by two tables

F	q_1	q_2	$\ldots$	q_1
s_1	r_3	r_1	$\ldots$	r_j
s_2	r_2	r_3	$\ldots$	r_h
.	.	.	$\ldots$	.
s_m	r_k	r_j	$\ldots$	r_i

G	q_1	q_2	$\ldots$	q_1
s_1	q_5	q_3	$\ldots$	q_j
s_2	q_1	q_2	$\ldots$	q_h
.	.	.	$\ldots$	.
s_m	q_k	q_j	$\ldots$	q_i

or equivalently by a state transition diagram:

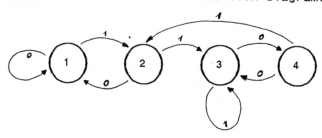

A conceptual implementation of a finite state machine is
shown in fig. 2. It consists of a Boolean function and a
set of latches. The Boolean function maps the combination of
the input S and the previous state Q onto an output R and a
new state Q.

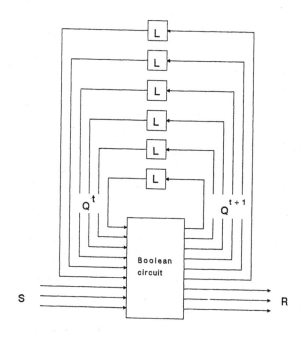

Fig. 2 A finite state machine

The complexity of a state machine is sufficiently large to
describe a computer. According to Fig. 3 any binary
computation can be considered as an iteration (path) in the

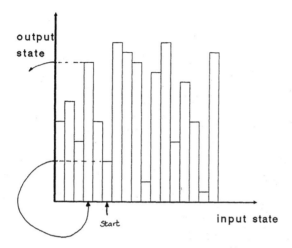

Fig. 3 Computation is an iteration in state space.

state space. The complexity of a state machine can be expressed by the number of possible transition tables depending on the number of states l, the number of inputs m and the number of outputs n:

$$C_{FSM} = 2^{(n+1)\,2^{(m+1)}}$$

A reasonable computer needs more than 1000 bits of memory and additional registers. Implemented as a finite state machine, the transition table for a computer would have to be larger than 2^{1000} bits. This makes it impractical to implement a computer as a state machine. The reason for this large complexity is that the most general state machine allows that an output may depend on all the inputs and previous states. In a practical situation the output depends only on a few number of input bits and state bits. The usual way around this problem is to divide the computer into units (von Neumann) which can be described as simpler finite state machines and to minimise the state transition function. Minimisation is a technique based on Boolean algebra that allows to express a Boolean function by less minterms.

4.

PROGRAMMABILITY IN A VON NEUMANN COMPUTER

The von Neumann computer[2] can be decomposed into four state machines: the arithmetic/logic unit (ALU), the control unit (CU), the memory (MEM) and the input-output unit (I/O). Fig. 4 shows how these units are connected (the I/O unit is omitted for simplicity). Fig. 5 a,b,c show the ALU, the CU and the memory as state machines. A and B are the registers for logic and arithmetic operations. The program counter P, as in the design of the von Neumann machine is included in the ALU. It has to be incremented for normal operations and loaded from memory for jumps. The condition input allows conditional jumps. Data can also be transferred between these registers. The control unit contains the instruction register I and a count cell C. The instruction register is loaded from the ALU. By incrementing the count cell, a sequence of micro

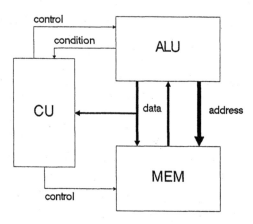

Fig. 4 Connection of the units of a von Neumann computer

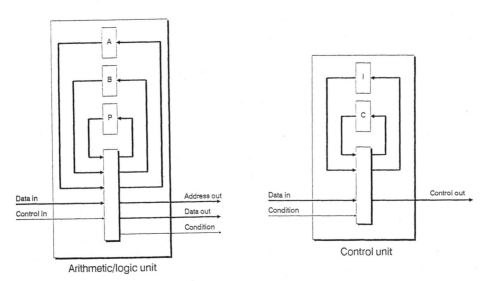

Arithmetic/logic unit

Control unit

Fig. 5 a) The ALU   b) The control unit as state machine

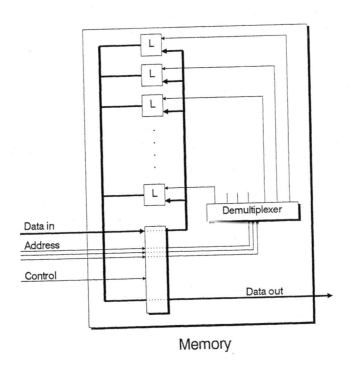

Memory

Fig. 5c The memory as a state machine

instructions can be executed. As a result a sequence of control information is generated at the output. The memory allows two functions: reading and writing of data. The address used for both functions is converted from binary to '1 of N' coding (demultiplexing). The demultiplexer is is actually part of the combinatorial logic function but is shown here separately for clarity.

5.
OPTICAL FINITE STATE MACHINES

Finite state machines ideally require a combinatorial logic function and a feedback loop containing latches. An overview over the optical implementations of combinatorial logic has been given in part II. Here we list some approaches to optical finite state machines as a basis for optical digital computers.

sequential optical logic[3]:

The feedback loop contains space variant interconnections and the gates provide combinatorial logic function between a pair of two inputs.

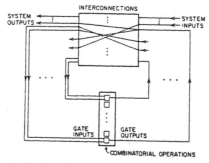

Fig. 1. Functional block diagram of sequential optical logic.

Fig. 6 K. Jenkins et al[3]

Optical parallel array logic system (OPALS):

A similar arrangement shows the architecture for OPALS[4]. Here however regular interconnections are used because the intended application is image processing. The combinatorial logic is realized with the method of shadow casting logic:

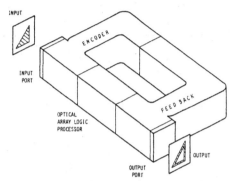

Fig. 2. Schematic diagram of the OPALS. An image to be processed is put into the input port of the system, coded by an encoder. and processed by an optical array logic processor. The result is put out from the output port of the system or fed back to the encoder.

Fig. 7 J. Tanida, Y. Ichioka [4]

Programmable processor based on symbolic substitution[5]:

In this design, the combinatorial logic block is decomposed into a sequence of blocks performing simple operations. The first two blocks (Fig. 8) are used to shift data within the two-dimensional array. The third block provides two logic functions (AND, XOR) and the identity operation. The subsequent switch block determines which of these logic functions will be used. Every function in each block is realized by (space invariant) substitution rules. The rules for the four blocks are shown in fig. 8 a,b,c. Space variance is introduced by the (space variant) control information entering at each stage exept the stage where logic is performed unconditionally.

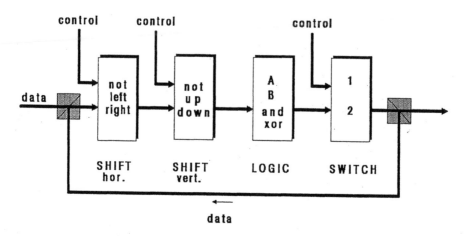

Fig. 8 Programmable processor based on symbolic substitution

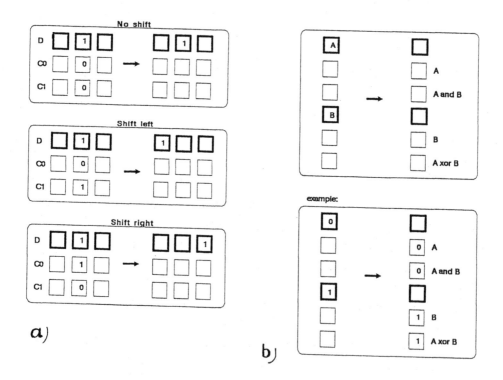

a)

b)

Fig. 9 Rules for the processor: a) rules for horizontal shift, b) rules for logic, c) rules for the select operation.

c)

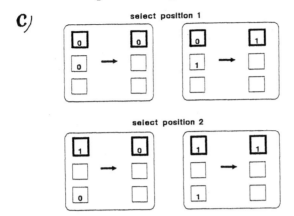

select position 1

select position 2

References:

1. see for example: M.L. Minsky: "Computation: finite and infinite machines", Prentice Hall, Inc., Englewood Cliffs, N.J. 1967

2. A.W. Burks, H.H. Goldstine, J.v.Neumann: US Army Ordnance Department Report, 1946. printed also in: C.G. Bell and A. Newell, "Computer structures: readings and examples", New York Mc Graw Hill 1971, p. 92-119.

3. B.K. Jenkins, P. Chavel et al.: "Architectural implications of a digital optical processor", Appl. Optics 23, 3465 (1984)

4. J. Tanida, Y. Ichioka: "Optical parallel array logic system. 2: A new system architecture without memory elements", Appl. Optics 25, 3751 (1986)

5. K.-H. Brenner: "Programmable optical processor based on symbolic substitution", Appl. Opt. 27, 1687 (1988)

PERCEPTRON LEARNING IN OPTICAL NEURAL COMPUTERS

David Brady and Demetri Psaltis

California Institute of Technology, Pasadena, CA 91125

1.

INTRODUCTION

A neural computer consists of simple nonlinear processing units, "neurons", and a large number of linear interconnections between neurons. Information is stored in the computer almost exclusively in the interconnection pattern. A neural computer is "programmed" by finding an interconnection pattern such that when the activities of the neurons in the network are initially represented by a vector $\vec{x}$ they evolve to a final state $\vec{y}$ which is associated with $\vec{x}$. One of the primary attractions of neural computation involves the existence of "learning" algorithms which can construct interconnection patterns which solve a given problem, i. e. find an appropriate interconnection pattern, from sequences of inputs $\vec{x}(n)$. If the output $\vec{y}(n)$ to be associated with $\vec{x}(n)$ is used to program the network the learning algorithm is "supervised". If the learning algorithm discovers $\vec{y}(n)$ by using more general constraints the algorithm is "unsupervised". In either case, the network is self-programming in the sense that the learning algorithm discovers an interconnection pattern which satisfies a given set of input-output constraints. In implementing a neural computer it is necessary to provide sufficient interconnections and neurons to address the problems to which the network will be applied and to allow for modification of the interconnections in order to search for the interconnection pattern which implements a desired transformation.

Optical holography is well suited to implementations of neural computers since it can provide a large number of interconnections relatively simply[1,2]. This is because globally broadcast optical signals may be superposed in free space without crosstalk. In this paper we consider the use of photorefractive holography for interconnections in neural computers[3,4,5,6,7]. Holograms are recorded in photorefractive materials via the electrooptic modulation of the index of refraction by space charge fields created by the displacement of photogenerated charge carriers[8,9,10]. Due to the 3-D nature of volume holograms, a large number of independent gratings may be stored in photorefractive crystals[11]. In order to use these crystals to implement a given neural model, we must find a way to arrange the charge in the crystal to correctly reflect the interconnection pattern required by the model. In section II of this paper we describe how holographic interconnections appropriate for optical neural computers may be stored in a photorefractive crystal. In section III we describe an adaptive optical architecture which uses dynamic holographic recording in a photorefractive crystal to implement a perceptron-like neural model.

2.
VOLUME HOLOGRAPHIC INTERCONNECTIONS

The state of a neural computer subject to a given learning algorithm is described by a matrix $\bar{\bar{w}}$ specifying the strengths of the connections between each pair of neurons and a vector $\vec{x}$ describing the activity of the neurons. In an optical system the activity of a single neuron may be represented by either the electrical field strength or the intensity in a given spatial mode. The interconnection weight between two neurons may be represented by the efficiency with which the field or the intensity is transferred from one mode to the other. A basic framework for an optical neural computer is shown in fig. 1. In this system the activity of a given neuron is represented by the field or intensity at a specified point in an input "neural plane". A volume hologram separated from the neural planes by a pair of Fourier lenses diffracts light emitted by neurons on the input plane onto detectors corresponding

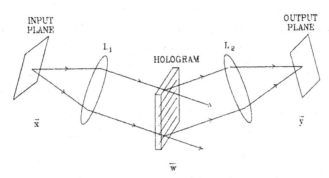

INPUT PLANE

OUTPUT PLANE

L_1

HOLOGRAM

L_2

$\bar{x}$

$\bar{y}$

$\bar{w}$

Fig. 1 Optical neural computer architecture.

to the input ports of the neurons on the output neural plane. The strength of the interconnection between a pair of neurons corresponds to the diffraction efficiency of the holographic grating which diffracts light from the input neuron onto the output neuron.

Three fundamental issues arise in implementing adaptive neural models based on the geometry of fig. 1. First, the number of degrees of freedom available for storage in the refractive index of the crystal is $O(N^3)$, which is less than the number of degrees of freedom needed to fully and independently interconnect every pair of points on the neural planes, $O(N^4)$. N is the number of resolvable spots along any one dimension of the optical system. Second, the number of degrees of freedom available to control the volume hologram in any single monochromatic exposure, $O(N^2)$, is less than the number of degrees of freedom which must be controlled in the volume, $O(N^3)$. And third, the mathematical models for learning which can be implemented using a given holographic medium are constrained by the physics of the holographic recording process. The first two problems are briefly reviewed here and the third is explored in the next section.

In the geometry of fig. 1, the discrepancy between the number of interconnections which can be stored in a volume hologram, N^3, and the number of connections which are possible between the two planes, N^4, gives rise to degeneracies; multiple pairs of input/output points are coupled by a given connection grating. Bragg phase matching constraints prevent shift invariance for shifts along the wavevector which couples a given pair of neurons. However, pairs which are shifted from a coupled pair perpendicular to the grating wavevector satisfy the Bragg condition for that grating and are also coupled. In [12,13,14], these degeneracies are analyzed and grids of appropriate fractal dimension are derived such that all points on the grids may be fully and independently interconnected. The number of neurons in each fractal grid for full interconnection may be N^{3-d} at the input neural plane and N^d at the output neural plane for $1 \geq d \leq 2$. Fractal grids of higher dimension with independent but local connections are also possible[15]. The use of such fractal grids resolves the problem of the mismatch between the dimension of the volume and the dimension of the interconnection matrix.

The implementation of $\bar{\bar{w}}$ using a volume hologram may be described by coupled wave equations in Fourier space. Consider the system of fig. 1 implemented with fractal grids of dimension $\frac{3}{2}$. The electric field $\vec{E}(\vec{r})$ propagating in the hologram may be represented as a weighted sum of plane waves emanating from the input neurons or incident on the output neurons.

$$\vec{E}(\vec{r}) = \sum_i A_i e^{j\vec{k}_i \cdot \vec{r}} \tag{1}$$

where A_i represents the field amplitude corresponding to the ith neuron. The effect on $\vec{E}(\vec{r})$ of the hologram recorded in the volume may be described by substituting eqn. (1) into the Helmholtz equation and applying the slowly varying envelope approximation. From this approach we obtain

$$jk_{iz}\frac{\partial A_i}{\partial z} = \sum_j \kappa_{ij} A_j \tag{2}$$

where k_{iz} is the component of the propagation vector $\vec{k}_i$ along the optical axis and κ_{ij} is the holographic perturbation spatial frequency $\vec{k}_i - \vec{k}_j$. Equation (2) may be rewritten as follows:

$$\frac{\partial \vec{A}}{\partial z} = \bar{\bar{w}}\vec{A} \tag{3}$$

where the ith component of $\vec{A}$ is A_i and $w_{ij} = -j\frac{\kappa_{ij}}{k_{iz}}$. The solution to eqn. (3) is $\vec{A} = exp(\bar{\bar{w}}z)\vec{A}_o$, where $\vec{A}_o$ represents the incident field. If all of the eigenvalues of $\bar{\bar{w}}$ satisfy $|\lambda|L \ll 1$, where L is the length of the holographic interaction region, then

$$\vec{A} \approx (\bar{\bar{I}} + \bar{\bar{w}}L)\vec{A}_o, \tag{4}$$

where $\bar{\bar{I}}$ is the identity matrix. If $\vec{A}_o$ is of the form

$$\vec{A}_o = \begin{pmatrix} \vec{x} \\ \vec{0} \end{pmatrix}, \tag{5}$$

then output field $\vec{A}$ consists of an undiffracted term $\bar{\bar{I}}\vec{x} = \vec{x}$ plus a diffracted signal term $L\bar{\bar{w}}\vec{x}$. Thus the hologram written in the volume performs a vector-matrix multiplication taking the input vector represented by plane wave components of the incident field onto an output vector represented by plane wave components of the diffracted field. The Fourier lenses in fig. 1 transform the arrays of plane wave to arrays pixels. Under suitable conditions of mutal incoherence between the stored gratings and the field components, vector-matrix multiplication can be shown to occur between the incident and output intensities.

The second issue mentioned above is the control of $\bar{\bar{w}}$ using $\vec{E}(\vec{r})$[17]. Assuming $M = N^{\frac{3}{2}}$ neurons in each neural plane, $\bar{\bar{w}}$ is an $M \times M$ matrix containing $M^2 = N^3$ degrees of freedom. $\vec{E}(\vec{r})$ consists of the sum of the activities of the neurons, which has at most $2M$ components. In order to specify each interconnection independently, the fields used to record $\bar{\bar{w}}$ must contain at least one degree of freedom for each interconnection. Since there are M^2 interconnections and only $2M$ recording beams, it is not possible to record $\bar{\bar{w}}$ in a single exposure. The degrees of freedom of the field may be increased, however, by making multiple exposures. In particular, in M exposures there are $2M^2$ degrees of freedom in the M recording fields. Even with enough degrees of freedom, however, whether or not M exposures is sufficient to specify $\bar{\bar{w}}$ depends on the dynamics of hologram recording. In media for which the holographic perturbation is linear in the recording intensity, that is where

$$\kappa_{ij} \sim \sum_n I_{ij}(n), \qquad (6)$$

where the sum is over exposures and $I_{ij}(n)$ is the ij^{th} Fourier component of the intensity at the nth exposure, it turns out that M exposures are sufficient to control $\bar{\bar{w}}$[17]. We describe below an exposure schedule for photorefractive crystals which allows the formation of holograms which are linear in the modulation terms of the writing intensity. Due to the erasure of previous exposures by the current exposure, the total diffraction efficiency of the multiply exposed hologram falls off. This fall-off effectively limits the number of exposures which can be made and thus the amount of information which can be stored in a photorefractive hologram.

A photorefractive hologram is created by a space charge distribution in the volume of an electrooptic material. The amplitude of this space charge distribution is limited by the photo-active trap density in the material. We have shown that, particularly for incoherent systems, the saturation of the space charge distribution limits the interconnection density in photorefractive crystals by constraining the mean square diffraction efficiency per interconnection when a large number of interconnections are recorded[4]. A second limitation on the interconnection density which may be recorded in photorefractive crystals arises from the fact that each exposure of the crystal causes a redistribution of the space charge. Thus if we wish to record an interconnection pattern as a series of exposures, each exposure partially erases previously recorded information. On the other hand, the fact that the charge in the crystal may be continually redistributed is exactly the property which makes photorefractive crystals attractive for adaptive systems since this means that the photorefractive response does not degrade under continuous recording.

The dynamics of the space charge distribution in photorefractive crystals may be specified by a set of transport equations describing the motion of photogenerated charge under the influence of the space charge field, applied fields, diffusion, and photovoltaic effects[8]. While the dynamics of the charge distribution may be quite complex, at low modulation depths in the absence of coupling between the writing beams the amplitude of the ηth Fourier component of the space charge field may be modeled in time by[10]

$$E_\eta(t) = E_{saturation} e^{-\frac{t}{\tau}} \int_0^t m_\eta(s) e^{j\phi_\eta(s)} e^{\frac{s}{\tau}} ds, \qquad (7)$$

where $m_\eta(t)$ and $\phi_\eta(t)$ are the modulation depth and phase of the ηth Fourier component of the intensity pattern in the crystal. τ is a characteristic time constant which is inversely proportional to the optical intensity. We assume that the power of the recording light is constant in time. In most photorefractive crystals, a hologram is created by the modulation of the index of refraction by the space charge field using the linear electro-optic effect. In this case, the time development of w_{ij} is described by

$$w_{ij}(t) = w_{sat} e^{-\frac{t}{\tau}} \int_0^t m_{ij}(s) e^{j\phi_{ij}(s)} e^{\frac{s}{\tau}} ds. \qquad (8)$$

If we make a series of exposures in which each m_{ij} is constant during each exposure then eqn. (8) may be written in discrete form as

$$w_{ij}(n) = w_{sat} \sum_{s=1}^{n} m_{ij}(s) e^{-\sum_{s'>s}^{n} \frac{t(s')}{\tau}} (1 - e^{-\frac{t(s)}{\tau}}), \qquad (9)$$

where $t(s)$ is the exposure time for the sth exposure and we have assumed for simplicity that $e^{j\phi_{ij}(s)} = 1$. $w_{ij}(n)$ is proportional to the sum of the m_{ij}'s if $e^{-\sum_{s'>s}^{n} \frac{t(s')}{\tau}}(1 - e^{-\frac{t(s)}{\tau}})$ is a constant for all s, which is the case if $t(1) \gg \tau$ and

$$t(s) = \tau \ln\left(\frac{s}{s-1}\right) \quad s > 1. \tag{10}$$

With these values for $t(s)$, eqn. (8) becomes[4,18]

$$w_{ij}(n) = \frac{w_{sat}}{n} \sum_{s=1}^{n} m_{ij}(s). \tag{11}$$

The total diffraction efficiency due to $\bar{\bar{w}}$ is given by

$$\eta = \frac{\sum_i |\sum_j w_{ij} A_j|^2}{|\vec{A}|^2} = \frac{w_{sat}^2 \sum_i |\sum_{j,s} m_{ij}(s) A_j|^2}{n|\vec{A}|^2}. \tag{12}$$

The rate at which η decreases with the number of exposures, n, depends on the statistics of $m_{ij}(s)A_j$. If this term is a random complex number of constant magnitude, a reasonable assumption for an associative network, the sum over s scales as $\sqrt{n}$ and η falls off inversely with n.

3.

A PHOTOREFRACTIVE PERCEPTRON

The perceptron[19,20] is an example of a supervised adaptive neural model which can be implemented using photorefractive crystals. A perceptron consists of a set of input neurons with activities described by a vector $\vec{x}$ which drive a single output neuron via a weight vector $\vec{w}$. The activity of the output vector is high if and only if $\vec{w} \cdot \vec{x} > \omega_o$ where ω_o is a fixed threshold level. A perceptron can be trained to separate a set of input vectors into two classes by various methods, the simplest of which involves updating the weight vector according to

$$\begin{aligned}
\vec{w}(n+1) &= \vec{w}(n) + \alpha\vec{x} \\
\omega_o(n+1) &= \omega_o(n) + \alpha
\end{aligned} \tag{13}$$

where $\vec{w}(n)$ is the state of the weight vector at discrete time n when $\vec{x}$ is presented for classification. α is zero if $\vec{x}$ is correctly classified and 1 (-1) if $\vec{x}$ is misclassified in the low (high) state. If training vectors from a set $\{\vec{x}\}$ are presented in sequence, eqn. (13) is known to converge on a weight vector implementing an arbitrary prescribed dichotomy if such a weight vector exists.

One means of implementing a perceptron in a photorefractive system would be to update each interconnection in series as prescribed by eqn. (13). This approach has two disadvantages. The first is that the saturable nature of the photorefractive response limits the range of w_i to $(-w_{sat}, w_{sat})$ for coherent systems and $(0, w_{sat})$ for incoherent systems. The weights determined by eqn. (13) may not be guaranteed to lie within these bounds. The second disadvantage to this approach is that each weight must be updated independently. In order to update the weights in this way we would need to detect the value of each weight and generate optical beams specifically to change that weight by the prescribed amount. In a volume holographic implementation with spatially multiplexed weights it is not possible to change the weights independently.

These problems can be avoided by implementing a perceptron in a photorefractive system using the dynamics of eqn. (9) to implement a variation of eqn. (13) which does not require us to update each connection with an independent exposure. This can be done particularly simply in an incoherent system. An architecture for an incoherent photorefractive perceptron[21] is shown schematically in fig. 2. Much of the complexity of the systems described in the previous section is avoided in this system so that we can concentrate on the use of photorefractive dynamics in learning. The input to the system, $\bar{x}$, corresponds to a two dimensional pattern recorded from a video monitor onto a liquid crystal light valve. The light valve transfers this pattern onto a laser beam. This beam is split into two paths which cross in a photorefractive crystal. The light propagating along each path is focused such that an image of the input pattern is formed on the crystal. The images along both paths are of the same size and are superposed on the crystal. The intensity diffracted from one of the two paths onto the other by a hologram stored in the crystal is isolated by a polarizer and spatially integrated by a single output detector. The thresholded output of this detector corresponds to the output of a perceptron. The fact that the connections in this system are stored locally in the image plane of the input allows us to very simply control each connection independently. This is at a cost, however, of the loss of the high connection densities achieved by using the entire volume of

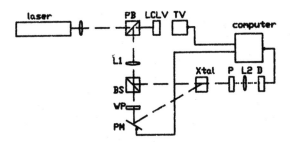

Fig. 2 Photorefractive perceptron. PB is a polarizing beam splitter.
L1 and L2 are imaging lenses. WP is a quarter waveplate. PM is
a piezoelectric mirror. P is a polarizer. D is a detector. Solid lines
show electronic control. Dashed lines show the optical path.

the storage medium in the Fourier domain.

The ith component of the input to this system corresponds to the intensity
in the ith pixel of the input pattern. The interconnection strength, w_i, between
the ith input and the output neuron corresponds to the diffraction efficiency of the
hologram taking one path into the other at the ith pixel of the image plane. In
analogy with eqn. (13), w_i may be updated by exposing the crystal with the input
along both paths. If the modulation depth between the light in the two paths is
high then where x_i is high w_i is increased. If the modulation depth is low between
the two paths then where x_i is high w_i is reduced. The modulation depth between
two optical beams can be adjusted by a variety of simple mechanisms. In fig. 2 we
choose to control $m(t)$ using a mirror mounted on a piezoelectric crystal. By varying
the frequency and the amplitude of oscillations in the piezoelectric crystal we can
electronically set both $m(t)$ and $\phi(t)$ over a continuous range without changing the
intensity in the optical beams or interrupting readout of the system.

We have implemented the architecture of fig. 2 using a SBN60:Ce crystal provided by the Rockwell International Science Center. We used the 488 nm line of an
argon ion laser to record holograms in this crystal. Most of the patterns considered
were laid out on 10×10 grids of pixels, thus allowing 100 input channels. Ultimately,
the number of channels which may be achieved using this architecture is limited by
the number of pixels which may be imaged onto the crystal with a depth of focus
sufficient to isolate each pixel along the length of the crystal.

Using the variation on the perceptron learning algorithm described below with

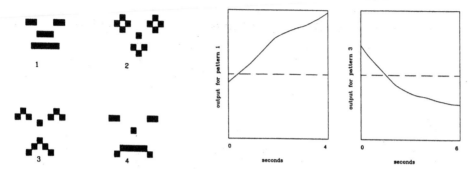

Fig. 3 Training patterns.   *Fig. 4 Detector output during training.*

fixed exposure times Δt_r and Δt_e for recording and erasing, we were able to correctly classify various sets of input patterns. An example of such a set is shown in fig. 3. In one training sequence, we grouped patterns 1 and 2 together with a high output and patterns 3 and 4 together with a low output. After all four patterns had been presented four times, the system gave the correct output for all patterns. The weights stored in the crystal were corrected seven times, four times by recording and three by erasing. Fig. 4a shows the output of the detector as pattern 1 is recorded in the second learning cycle. The dashed line in this figure corresponds to the threshold level. Fig. 4b shows the output of the detector as pattern 3 is erased in the second learning cycle.

In this system, eqn. (9) becomes

$$w_i(n) = w_{sat} |\sum_{s=1}^{n} m_i(s) e^{-\sum_{s'>s}^{n} \frac{t(s')}{\tau}} (1 - e^{-\frac{t(s)}{\tau}})|^2 \tag{14}$$

Two problems prevent the use of the exposure schedule of eqn. (10) in this system. The first is that the assumption of approximately constant intensity in each exposure is violated in an incoherent image plane system. The second is that, while the perceptron algorithm is known to converge, the number of training steps needed to reach convergence can be very large. If the exposure schedule were followed, the total diffraction efficiency after a training sequence could be very low. This prospect is particularly troubling since we know that not all exposures in a learning sequence contribute equally to the ultimate ability of the system to learn to solve a given problem. Indeed, the point in making a network adaptive is to find a comparatively succinct representation of a solution to a classification problem from a great deal of redundant input data.

In lieu of using an exposure schedule in our simple perceptron we assume that $t(s) = \Delta t_e$ in cycles in which erasure occurs and $t(s) = \Delta t_r$ in cycles in which writing occurs. Since τ is inversely proportional to the optical intensity, we can express $\frac{1}{\tau}$ at each pixel of the input as αx_i. Letting $m_i(t) = 0$ when the nth training vector yields too high an output, we find from eqn. (14) that the ith component of the weight vector is updated according to

$$w_i(n+1) = e^{-2\Delta t_e \alpha x_i} w_i(n) \tag{15}$$

Let $m_i(t) = m$ when the nth training vector yields too low an output, we find that the the ith component of $n + 1^{th}$ weight vector is

$$
\begin{aligned}
w_i(n+1) =& e^{-2\Delta t_r \alpha x_i} w_i(n) + m^2 w_{sat}(1 - e^{-\Delta t_r \alpha x_i})^2 \\
&+ 2m\sqrt{w_i(n)w_{sat}} e^{-\Delta t_r \alpha x_i}(1 - e^{-\Delta t_r \alpha x_i})
\end{aligned}
\tag{16}
$$

Note that, as one might intuitively wish, w_i decreases monotonically with x_i if the output for $\vec{x}$ is too high and increases monotonically with x_i if the output for $\vec{x}$ is too low. Eqns. (15) and (16) may be greatly simplified if we assume that x_i takes on only the values 0 and 1. Since the output power incident on the detector may be arbitrarily renormalized, we may also assume without loss of generality that $m^2 w_{sat} = 1$. Under these assumptions eqn. (15) may be recast in the form

$$\Delta w_i(n) = w_i(n+1) - w_i(n) = -(1 - \gamma_e^2) x_i w_i \tag{17}$$

where $\gamma_e = e^{-\Delta t_e \alpha}$. Eqn. (16) becomes

$$\Delta w_i(n) = -(1 - \gamma_r^2) x_i w_i + 2\gamma_r(1 - \gamma_r) x_i \sqrt{w_i} + (1 - \gamma_r)^2 x_i \tag{18}$$

where $\gamma_r = e^{-\Delta t_r \alpha}$.

We assumed both in our simulations and in our experiments that an acceptable value for ω_o could be guessed a priori. This assumption is unnecessary if one of the stored weights is used as a threshold. An example of convergence in simulations run under this training algorithm is shown in fig. 5. In this figure we plot $\delta \times (\vec{w}(n) \cdot \vec{x}^{(n)} - \omega_o)$ as a function of n. δ is 1 (-1) if $\vec{x}^{(n)}$ is in Ω_+ (Ω_-). The system has converged when a cycle through all stored vectors yields all positive outputs. This example involves ten randomly selected and classified training vectors drawn from a hundred dimensional space. Ω_+ and Ω_- each contain five vectors. The training vectors are presented twelve times in sequence before a solution vector was found. In the thirteenth sequence through the vectors $\delta(\vec{w}(n) \cdot \vec{x}^{(n)} - \omega_o) > 0$ for all n, indicating that a solution has been obtained. γ was 0.9 and the weights were randomly initialized. A learning curve under similar circumstances in the experimental system is shown in fig. 6. In this case nineteen cycles were necessary to converge on a solution vector. Most of the discrepancy between the experimental system and the simulations arises from the difficulty involved setting the learning parameters and initial conditions to be identical. The key trend to notice in both figures is that the amplitude and frequency of negative outputs decreases with each cycle´until convergence is reached. We found that it was relatively easy to achieve convergence both in simulations and in the experimental system.

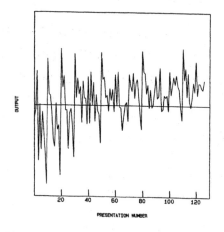

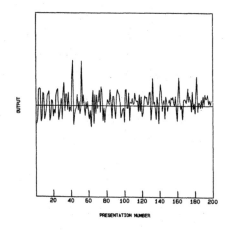

Fig. 5 Learning curve from simulations.   *Fig. 6 Experimental learning curve.*

A final point to notice about fig. 5 and fig. 6 is that convergence is achieved even though the exposure involved in each training cycle is large. We know from eqn. (14) that the effect of a given exposure on the final state of w_i is decreased by at least a factor of γ^q, where q is the number of exposures made after the exposure of interest. Since $\gamma^q \ll 1$ for $q > 10$, information recorded early in the training sequences is erased from the crystal before convergence is reached. The early exposures drive the system toward a region of weight space from which convergence is achieved, rather than store information.

4.

CONCLUSION

Photorefractive crystals hold the promise of very high density reconfigurable interconnections for neural computers. In order to realize this potential methods must be found to reconcile the dynamics of neural models with the physics of the photorefractive devices. The rudimentary perceptron algorithm described here is interesting because it relies directly on the dynamics of the photorefractive response. Since the dimension of the input can be fairly large, this algorithm may be of use in certain applications. However, in order to utilize all the capabilities of photorefractive crystals in optical neural computers, it will ultimately be necessary to design architectures which dynamically address the full volume capacity of the crystals.

ACKNOWLEDGEMENTS

The authors thank Ratnakar Neurgaonkar and Rockwell International for supplying the SBN crystal used in our experiments and thank Xiang-Guang Gu and R. Scott Hudson for their comments on various topics discussed here.

This research is supported by the Defense Advanced Research Projects Agency, the Army Research Office, and the Air Force Office of Scientific Research.

REFERENCES

[1] D. Psaltis and N. H. Farhat, Opt. Lett., **10**, 98 (1985).

[2] Y. S. Abu-Mostafa and D. Psaltis, Scientific American, pp.88-95, March, 1987.

[3] K. Wagner and D. Psaltis, Appl. Opt., 26(23), pp.5061-5076(1987).

[4] D. Psaltis, D. Brady, and K. Wagner, Appl. Opt., May 1988.

[5] Y. Owechko, Appl. Opt., 26(23), pp. 5104-5111 (1987).

[6] D. Z. Anderson, Opt. Lett., **11**,45 (1986).

[7] A. Yariv and S. K. Kwong, Opt. Lett., **11**,482 (1986).

[8] N. V. Kuktarev, V. B. Markov, S. G. Odulov, M. S. Soskin, and V. L. Vinetskii, Ferroelectrics, **22**,949(1979).

[9] J. Feinberg, D. Heiman, A. R. Tanguay, and R. W. Hellwarth, J. Appl. Phys. **51**,1297(1980).

[10] A review of the photorefractive effect is presented by T. J. Hall, R. Jaura, L. M. Connors, P. D. Foote, Prog. Quan. Electr. **10**,77(1985).

[11] P. J. Van Heerden, Appl. Opt., **2**,(4),393(1963).

[12] D. Psaltis, J. Yu, X. G. Gu, and H. Lee, Second Topical Meeting on Optical Computing, Incline Village, Nevada, March 16-18,1987.

[13] H. Lee, X. Gu, and D. Psaltis, "Volume holographic interconnections with maximal capacity and minimal crosstalk", to appear in J. Appl. Phys..

[14] D. Psaltis, X. G. Gu, and D. J. Brady, SPIE Proceedings **963**, 70(1988).

[15] X. Gu and D. Psaltis, "Local and asymmetric interconnections using volume holograms", OSA Annual Meeting, *1988 Technical Digest Series*, **11**, Optical Society of America, Washington, DC, 148(1988).

[16] H. Kogelnik, "Coupled Wave Theory for Thick Hologram Gratings", Bell Sys. Tech. Journal, **48**(9),2909(1969).

[17] S. Hudson, D. J. Brady, and D. Psaltis, "Properties of 3-D imaging systems", OSA Annual Meeting, *1988 Technical Digest Series*, **11**, Optical Society of America, Washington, DC, 74(1988).

[18] K. Bløtekjaer, Appl. Opt., **18**, 57(1979).

[19] F. Rosenblatt, Principles of Neurodynamics: Perceptron and the Theory of Brain Mechanisms, Spartan Books, Washington,(1961).

[20] M. L. Minsky and S. A. Papert, *Perceptrons*, MIT, Cambridge, 1988.

[21] D. J. Brady, X. G. Gu, and D. Psaltis, SPIE Proceedings, **882**, 20(1988).

Current Technology Trends

K. H. Bohle

IBM Germany, Stuttgart

1 Introduction

Two developments have shaped the electronics industry: the invention of the transistor in 1947 and the photolithographic process developed around 1960. The advantages of the transistor were so significant that a short period of less than 12 years was sufficient to convert a laboratory curiosity into a mass product. The second invention was not so spectacular for the outside observer, but probably more important to the electronics industry. With photolithography it is possible to produce a large number of similar or identical components with a single sequence of production steps in a single piece of silicon. This manufacturing process eliminates the linear relation between the number of components and product cost.

The production cost of integrated circuits is proportional to the number of processing steps and the processed area of the silicon surface, but is only marginally dependent on the total number of electronic components integrated. Every reduction in size of the transistors results automatically in the reduction of cost per transistor.

If we try to find a period in the history of technology with a similar progress, we have to go back to the 15th century. The invention of mechanical printing cut a one-to-one relation: *One book = one human life.*

Up to the time of Gutenberg, one monk was able to copy one book (or a very small number of books at the most) per lifetime. With the invention of mechanical printing, information distribution was suddenly available for nearly nothing. An unprecedented social change was the result.

Microelectronics and microprocessors in particular give us 'information processing' for nearly nothing. If the progress continues at the same rate as observed during the past twenty years, we should be prepared for something spectacular.

In order to get some feeling for the future developments, we will use a very simple technique frequently applied by engineers:

- establish the state-of-art

- evaluate the rate-of-change

- observe external constraints

- look for absolute limits

2   State-of-the-art products

The most significant application for microelectronics is the digital computer. The circuits used for digital computers are all based on *very-large scale integrated (VLSI)* silicon chips. The state-of-the-art in VLSI electronics is nicely described in the very influential book by Carver Mead and Lynn Conway [20]. Chip size varies depending on the application from a few mm^2 to $1cm^2$. The production unit is the wafer. The wafer diameter has increased from 12 mm only 25 years ago to 200 mm at present. The smallest dimensions are in the order of $1\mu m$. Masks are now frequently produced by electron beam techniques, but optical projection lithography is still used for the chips. The impurity profiles are produced by diffusion and ion-implantation.

2.1   Memories

A *random access memory (RAM)* is a two-dimensional array of storage cells. Each cell stores one bit of information and is identified (selected) by an address. The address is usually a binary number. In order to write into or read from the cell, the address must be present at the addressing lines. The sequence of addresses can follow any pattern and is usually not constrained in any way (random).

2.1.1   Dynamic Memories (DRAM)

In a dynamic RAM, the storage cells are small capacitors. The information is stored as a charge on the capacitor. Each cell has a transistor associated for selection by the address-decoder.

The capacitor will slowly discharge due to leakage currents. A regular refresh is therefore required, in order to keep up the operation. Refresh cycle times of 1 to 10 ms are typical. The parameters of the first 1-Mbit chip in mass production [8] (since 1987) are listed in table 1.

2.1.2   Static Memories (SRAM)

In static memory chips each storage cell is made up of a pair of cross-coupled transistors. This circuit has two stable states. A third transistor is needed for the selection of the cell. This means, that a static cell is significantly larger than a dynamic cell

Type		DRAM	SRAM	μP	
Technology		NMOS	CMOS	NMOS	
Datapath		1, 2, or 4	8	32	bit
Capacity		1	1		Mbit
Size		5.5×10.5	8.0×13.65	8.4×8.4	mm^2
Cell size		4.1×8.8	6.4×11.6		μm^2
Access time	t_{acc}	40	35		ns
Cycle time	t_c	65	100	33	ns
Power	(active)	0.625	0.1	10	W
Power	(standby)	50	0.025		mW
Refresh		4	na	na	ms
Gate oxide	T_{ox}	25 (15)	20	25	nm
Photolitho.	minimum	1.0	1.0	1.0	μm
	average	1.5	1		μm
	Metal 1 width	1.2	1.2	1.5	μm
	Metal 1 space	1.0	1.0	1.0	μm
	Metal 2 width	na	1.6	5.0	μm
	Metal 2 space	na	1.2	3.0	μm
	Poly 1 width	1.0	1.0	1.5	μm
	Poly 1 space	1.0	1.0	1.0	μm
	Poly 2 width	1.0	1.0	na	μm
	Poly 2 space	1.2	1.2	na	μm
Pins		32	32	251	

Table 1: Typical VLSI Product Parameters

and therefore more expensive. With the same production technique, a SRAM chip has a smaller capacity. No refresh is necessary and the device keeps the information as long as the supply voltage is present. For many applications it is desirable, that the informations is even retained after a power failure has occured. This is not difficult to achieve with a static cell, because only a small current is required. This current can come from a small battery integrated into the second level package, usually a printed circuit board (PCB). Most static memories have short cycle times and simpler timing conditions because no refresh has to take place. The parameters of a chip not yet in mass production [10] are found in table 1.

2.2 Processors

In order to build a computer, we need the following building blocks:

- input

- output

- control

- arithmetic and logic unit (ALU)

- memory

For technical reasons, the control and ALU are frequntly combined in a building block called *processor* or *engine*. The processor was originally made up of a large number of components.

2.2.1 Microprocessors (μP)

Soon after the production processes for semiconductor memories were available in 1971, engineers put an entire von-Neumann-Processor on a single chip, and called it a *microprocessor*. The first microprocessor operated only on four bit in parallel. The standard chips of today have a datapath of 32 bits. The parameters of a typical chip [9] are found in table 1.

2.3 Application-Specific Integrated Circuits (ASIC)

For certain applications, the computing power of a microprocessor is not sufficient. These more powerful processors are than made up of chips dedicated to specific operations. They are called *application-specific IC's* and a large number of different chips (part numbers) must be designed. If all were designed individually (custom-design), the cost of the product would be very high. Many different techniques are used to keep the design of these chips at an affordable level.

- Gate Arrays

- Macro Library

Technology		Bipolar	CMOS	CMOS	
Photolitho.		2.5	1.0	1.25	μm
Size		7.4×7.4	9.4×9.4	3×3	mm^2
Cells	(wirable)	7250	27720		
Circuits	(2W equ.)	14000	40000	6000	
Delay	(worst case)	1.4	2.1	0.9	ns
Power	(per ckt)	0.54	0.1		mW
	(per Chip)	7	1.5	0.3	W
Interconnections	layers	4 metal	2 metal	2 metal	
Metal pitch		6.6	3.3		μm
Pins		240	231	124	

Table 2: Application-specific IC's

- Master slice

- Semi-custom design

The technology parameters of a few chips [6,2] are summariesed in table 2

2.4 Packaging

Microprocessors and other integrated circuits have improved information processing significantly, but their computing power is limited. For large applications, many chips have to work together in one much more powerful processor. This interconnection of chips has one very important drawback: The traveling time of electrical signals on the wiring between the chips reduces the speed. Modern electronic circuits perform a logical operation in less than one nanosecond. In one nanosecond, light travels 30 cm and electrical signals on wires travel at approximately half the speed of light. Only 15 cm of interconnecting wiring have therefore the same effect on speed as the electronic circuits themselves.

In recent years, engineers have therefore concentrated on the reduction of wireing length. With the help of multilayer ceramic substrates and the unique soldering connection (controlled collapse chip connection = C4) between the chips and the substrates, engineers have been very successful.

As an example, I will use the IBM Thermal Conduction Module (TCM) technology. On the modules only 90 mm x 90 mm in size, 133 chips are mounted. Each chip is connected through 121 contact pads (C4) to the substrate. This gives a total of more than 16,000 connections between chips and module, all produced by a single heat applying process. During operation, the chips produce more than 3 Watt of heat per cm^2. This heat is carried to the cooling liquid by the metal structure of the TCM. Due to the low thermal resistance between chip and cooling liquid, the junction temperature of the transistors is considerably lower than in air cooled modules. The reliability of components goes down exponentially with the junction temperature.

Parameter	Physical Unit	Technical Unit
propagation time	s	ns
clock frequency	Hz	MHz
power consumption	W	mW
size	m^3	μm^2
production cost		\$, £, Yen, DM ...
weight	kg	mg
reliability (MTBF)	s	% /kh
fan-in		1
fan-out		1
noise-immunity		%

Table 3: Selection parameters

3 Trends

3.1 Evaluation Criteria

In order to evaluate different technologies, we must understand the parameters that are used to select a particular product. The most important selection parameters are listed in table 3. The speed of the machine depends on the propagation time and the clock frequency is dictated mostly by this parameter. For logic circuits it is the time at which the logical result is available after the input signals are applied. For memory circuits it is the access time, at which the data is available after the address is applied. For certain pipelined structures, this parameter is less important than the clock frequency, but for a general purpose computer, the propagation delay is by far the most important parameter.

The power consumption has two influences: the packaging must be able to supply the circuits with the necessary currents und the produced heat must be carried away. The product of propagation-delay and power consumption is often called the *speed-power-product*. This parameter is a good figure-of-merit for simple comparisons. It is the energy required to perform a certain function.

The parameter that has the biggest impact on speed is the size. Production cost and weight are strongly related to size.

Reliability is measured as *mean-time between failures (MTBF)*. It has therefore the dimension of time, but engineers measure it usually as *% per 1000 hours (% /kh)*.

The logic circuits can have only a limited number of inputs and can only drive a limited number of other circuits. Engineers refer to these figures as *fan-in* and *fan-out*.

The susceptibility to internal as well as external influences is measured as noise immunity, expressed as a certain percentage of the information carrying commodity (e.g. voltage).

3.2 Integration

Recent history has shown, that most parameters improve, if size is reduced. Integration has been the name of the game. If we look at memory chips, we see that for DRAM, integration has come from 1 kbit per chip in 1970 to 1 Mbit in 1986. This is approximately a factor of 100 in a decade. If this trend continues, we would have between 256 Mbit and 1 Gbit per chip by the turn of the century. Is this likely? At this point I shall quote Gordon Moore (Founder and President of Intel):

> Any time you make extrapolations that depict straight lines on log paper, you come up with a catastrophe. But the main utility of performing the task is to show what kind of catastrophe to look for so that you can start soon enough to make the necessary adjustments to live with it.

4 Hurdles

There are principally two classes of 'catastrophes' to live with. The most serious class is governed by the basic laws of nature. We will look at these later.
In addition to these fundamental limits, with which we have to live in any case, a number of 'technical limits' have to be removed before significant progress is possible [23]. Some of the known hurdles are:

- Scaling with size reduction

- Wavelength limit for photolithography

- Heat removal

- Sensitivity to radiation

- Complexity in design

- Duration of functional test

For most of these problems, solutions are already within sight. Not all parameters change by the same amount, if we reduce the dimensions by a constant factor. Fortunately, some technologies rely on ratios rather than absolute values and therefore do not suffer that much from scaling problems. The CMOS (Complementary MOS) technology has profited from this in the past and will continue to do so in the foreseeable future. The CMOS circuits are slow, because the gate capacitors have to be charged through large resistors. Size reductions will reduce the capacitances significantly without increasing the impedances. To overcome the remaining limitations, lower operating voltages (1 V rather than 5 V and lower operating temperatures (77 K) may be required.
In order to produce dimensions smaller than the wavelength of light the optical exposure in photolithography must be replaced by something with a shorter wavelength. Electron beams, X-rays, and the synchrotron radiation (soft X-rays) are possible candidates. Electron beam lithography is well established, but it is a sequential process

and therefore expensive. Parallel E-beam techniques have been developed [7] but the progress in synchrotron radiation lithography looks very promising.

The heat produced by the individual circuits has been reduced steadly for twenty years, but so has the reduction in size. The problem today is that we have to take a small amount of heat out of very small volumes. Air cooling is not sufficient and even liquid cooling is not too good because of the boundary between the device and the liquid. Certain modifications to the surface [24] will be required.

Heat removal becomes much simpler, if the circuits are operated at reduced temperatures. With liquified gasses (helium or nitrogen) heat can be removed more easily and the amount of heat generated can be reduced significantly.

With the introduction of the 64 kbit DRAM chip it was found that α-particles caused *soft errors*. The particles discharged the capacitors holding the information, but caused no permanent damage. The reason is, that in such a circuit, all components are in high impedance state and therefore susceptable to radiation. For improved radiation immunity, static CMOS circuits have been developed that can cope with this because in such a circuit, the controlling device is in a low impedance (conducting) state.

There are quite a few other hopes for higher integration and it is not unlikely, that we will see 256 Mbit (or even 1 Gbit) chips by the year 2000.

4.1 Design and Test

The design of a memory chip is a relatively easy task: a large number of identical cells is arranged in a regular matrix. The design complexity is small in spite of the large number of components. With logic circuits and microprocessors, it is very different. Since 1971 the number of gates per chip has increased from 700 to approximately 100000. If this trend continues, we would have some 5 to 10 Million gates per chip by the year 2000. We may be able to produce chips with that many gates but can we cope with the complexities of design and test? The complexity related to design and test of the circuits grows very rapidly. It is generally accepted, that in the near future the emphasis of research will have to shift from the classical manufacturing problems to the management of complexity.

Experience with software engineering has shown that human beings can cope only with limited complexity. A typical software design team will be faced with one design error within 1000 lines of code. In many respects, one line-of-code (LOC) is equivalent to one logic gate on a chip and we find the same relative number of errors on VLSI chips as in software. There is just one big diffence: it is much more difficult to rectify a design error on a chip because it requires a redesign of the masks. The proposed remedies are either a very *regular processor structure* or an efficient *silicon compiler*. The successful solution will probably be a combination of the two: a powerful design environment (CAE=computer aided engineering) used for the (error-free) design of highly regular products.

5 Alternatives

5.1 Gallium Arsenide

For many years, alternatives for silicon have been investigated. The most popular alternative semiconductor is gallium arsenide.

Parameter	Si	GaAs	
Electron mobility	1400	8500	$cm^2/V\,s$
Substrate resistance	< 100	$> 10^6$	Ωcm
Bandgap	1.12	1.43	eV

The advantages of GaAs are:

- high electron mobility

- high substrate resistivity

- wide bandgap

- direct bandgap

- radiation hardness

- low noise

The advantages of Silicon look small:

- native oxide

- "natural"

- high quality single crystals

- low cost

A careful analysis will show, that the advantages of GaAs are only important for a small fraction of the circuits used in computers. The application areas for GaAs come mostly from the border between electronics and optics. The direct bandgap property of GaAs is giving the superiority. For normal computer circuits, the lack of the native oxide and the difficult production process does not allow the same level of integration for GaAs as compared to Si. The chips have to be smaller and the circuits are bigger. A large fraction of the cycletime of a processor comes from the travelling time of the signals between parts of the system. A system less highly integrated will therefore be bigger and the travelling times will be higher and the cycle time will be longer. Extrapolations have shown, that a GaAs Processor may be twice as fast as a processor based on Si. This small margin will hardly justify the investments for a new technology.

5.2   Superconductivity

For many years engineers have tried to employ superconductivity to build computers. In the 1950's the *Buck-Cryotron* looked very promising and in the 1970's the
Josephson-Junction inspired many. Both technologies were superior to the leading
technology of the day, but to such a small margin that it did not justify the changes
necessary.

With the hot-superconductors based on the *Zürich-Oxides* the discussion has started
all over again. If a computer is immersed in in liquid nitrogen (77 K), many advantages can be used:

- good cooling

- small temperature gradients within system

- superconducting wires

- high substrate resistivity

Time will show, if superconductivity will make it this time.

5.3   Wafer-Scale Integration

The production unit for integrated circuits is a wafer containing many chips. To
produce a system, many chips have to be interconnected. Why can't we just build a
system on a single wafer? The reason is a matter of yield. The bigger the chip, the
higher the probability for a bad spot and therefore a malfunctioning circuit. To have
an acceptable yield, wafer-scale integration requires a fault-tolerant logical structure.
This overhead has in the past been more expensive than the packaging of many chips
on a carrier. Impovements in production techniques may change this.

5.4   Three Dimensional Structures

All integrated circuits in production today are two dimensional structures. To pack
many layers of these on top of each other is very appealing. Up to ten layers have
been proposed and even produced in the laboratories [1]. If this development will
lead to mass product will be seen in a couple of years. The prospects are very good,
because this will reduce the problems of wireing and the average wire length is smaller
which will speed up the product.

6   Structures

Although mathematicians and computer scientists have invented a large number of
different computer structures, nearly all computers in production today are of the
von-Neumann type [5].

It is the result of one hundred years of research in calculating machines [22]. The
von-Neumann computer was designed for problems in numerical mathematics. Most

computers today are used for very different applications, but all attempts to change have failed so far.

This can be explained partly by the investments made for this structure and therefore compatibility has become the most significant goal. *Non-von-Neumann* structures will have to be orders of magnitude better (whatever that means) in order to be accepted. This is very unlikely for the applications of today, because the von-Neumann computer can cope with these.

Completely new applications will require not only new computer structures, but also new interfaces to the user. In recent years, applications have surfaced, that will result in new structures and applications [18,14] Whereas classical applications were mainly concerned with information storage and computing, future applications will include logical reasoning, pattern matching and some sort of learning. The first step in that direction can be found with relational data bases and expert systems. For these applications, the present computers are still usable, but new architectures are looking advantageous. For the large class of knowledge based systems which are under consideration a new approach is very likely.

7 Limits

The literature on the limits of microelectronics and computing machines is growing very rapidly:[15,4,11] The fundamental components necessary to build information processing machines are storage elements and logic functions. Storage elements can be reduced to logic functions with feedback as seen in the cross-coupled NOR circuit. Mathematicians have invented structures with many values for logic variables (multi-valued logic, fuzzy sets etc.) but they are all based on an elementary *boolean algebra* operating on a set with just two elements: *true* and *false* or 0 and 1. We are therefore looking for storage elements with two stable states and functions to connect two variables in all ways. There are 16 different functions for the connection of two boolean values, but a subset can be found which is capable of generating the missing ones. The most popular *complete set of boolean functions* is the set containing the Functions *AND, OR,* and *INVERT*. Another complete set is formed by the NOR function alone. This is very important, because we can prove that we can build a computer with a new technology, if we have been able to implement the NOR function. It is important to recognize that a complete set of logic functions containes at least one irreversible function. This is not strictly true in the mathematical sense as research in *conservative logic* has shown, but it is still the case for all practical intents and purposes.

For the realization of computing machines we have to represent the logical variables by physical quantities like voltages, currents or pressures. We will call this physical quantity: g.

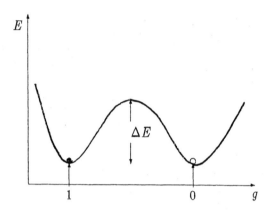

1: Bistable device

7.1   Thermodynamical limit

Any bistable storage device must have at least two energy minima in order to be stable as indicated in figure 1. The two minima don't have to be at the same energy level, but the assumption simplifies the argument. The energy hill (ΔE) between the two stable states is a measure of the reliability of the storage device. At temperature T the hill should be one or two orders of magnitude higher than the thermal energy kT. If we want to reset the cell from an unknown state into a known state (reset to "0"), we must invest at least ΔE. In theory, we could take this energy out again when "going downhill", but that is very cumbersome. A small amount $E_0 = ln2 \times kT$ cannot be recovered as shown by Landauer [19,12]. During the destruction process one degree of freedom in the information domain is lost, but one degree of freedom in thermodynamics is produced. We have therefore a minimal heat generation of $ln2 \times kT$ for every bit destroyed. When this was discovered in the early 1960's, hardly anyone bothered, technology was so far from the limit. Today we have travelled already halfway the distance. If we carry on as in the past, we will experience the limit in about twenty years.

The temperature becomes an important parameter if low energy dissipation is required. At roomtemperature (300K) $kT = 4 \times 10^{-21} J$. Reducing the operating temperature to the boiling temperature of liquid nitrogen (77 K) reduces the value by a factor of 4 and going to liquid helium (4.2 K) will give $kT = 5.52 \times 10^{-23} J$.

7.2   Quantummechanical limit

Suppose the device as shown in figure 1 has a switching time of Δt. If

$$\Delta E \times \Delta t < \hbar$$

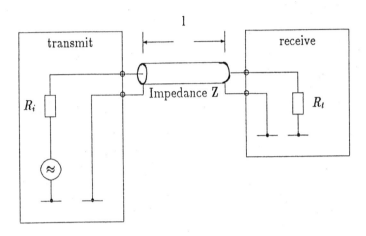

2: Transmission line

quantummechanical tunneling will take place. The result is an unreliable storage device. At least for deterministic computers we should avoid any technology that could come close to this limit.

7.3 Interconnection limit

The first two limits were fundamental in the way, that they apply to any kind of technology: electronic, optical, chemical, biological or else. The interconnection limit is well known for electronics but there are only strong indications that the limit is applicable to the other techniques as well. If we have two building blocks a distance l apart, we need a transmission line to communicate. If the time between two logic values (transition time, rise-time) is shorter than the travelling time on the line, we have to terminate the line by a resistor equivalent to the characteristic impedance Z of the line as indicated in figur 2.

7.4 Conservative logic

The absolute physical limits were investigated 15 to 20 years ago. There is an extensive literature on the subject. [19,16,12,17]
It had been observed from the very beginning of the investigations, that the irreversible nature of computation was a cornerstone of the theory. Charles Bennet [3] had frequently shown, that reversible computers are theortically possible. By 1982 Edmund Fredkin showed, that *reversible logic gates* exist. [13].

8   Conclusions

We have observed during the past 20 years the emergence of many new industries: microelectronics, telecommunications, and computers. Some observers believe, that the rapid changes are over. This is probably wishful thinking. We have no indications, that any slowdown will come. There are more signs for an accelleration of progress. In addition to the technological changes, we should be prepared for a change of society similar to the changes around 1500. Although we know, that anyone who predicts the future is either very brave or a fool, technology forecasting has become a significant discipline if not an industry in its own rights. In microelectronics and in computers in particular, predictions have been very misleading. I remember a discussion in 1968 about the prospects of of semiconductor memories as related to the than dominationg ferrite core memories. One semiconductor expert, who had build the most advanced integrated circuits that year remarked: *We will never be able to put 1000 Transistors in a chip and all transistors are within specification!* Two years later, the first 1 kbit semiconductor DRAM chip (intel 1103) was commercially available and the time of ferrite core memories had gone. On the other hand, we have seen many very promising new technologies that never made it:

- cryotrons

- parametric gates

- tunnel diodes

- thin magnetic films

- magnetic bubbles

- charge coupled devices

- Josephson junctions

That there is more to the success of a new technology than can be seen from the technical parameters, has been investigated by Pugh [21]. In additon, one element of surprise will never disappear: revolutionary changes can not be predicted by any methode.

References

[1] Yoichi Akasaka.   Three-dimensional IC trends.   *Proceedings of the IEEE,* 74(12):1703–1714, 1986.

[2] Floyd Anderson, Rick Brzozwy, and Sandy Metzgar. A 6k array with self-test and maintenance. *IEEE International Solid-State Circuits Conference,* ISSCC 87:150–151, 1987.

[3] Charles H. Bennett. Notes on the history of reversible computation. *IBM Journal of Research and Development*, 32(1):16–23, 1988.

[4] Charles H. Bennett and Rolf Landauer. The fundamental physical limits of computation. *Scientific American*, 38–46, 7 1985.

[5] Arthur W. Burks, Herman H. Goldstine, and John von Neumann. Preliminary discussion of the logical design of an electronic computing instrument. *Report to the U.S. Army Ordnance Department*, 1946.

[6] Delbert R. Cecchi and Robert F. Lembach. *IBM AS/400: System Processor Technology*. IBM Corporation, Rochester, 1988.

[7] Harald Bohlen et al. Electron-beam proximity printing. *IBM Journal of Research and Development*, 26(5):525–644, 1982.

[8] Howard L. Kalter et al. An experimental 80-ns 1-Mb DRAM with fast page operation. *IEEE Journal of Solid-State Circuits*, SC-20(5):914–922, 1985.

[9] Jeffrey Yetter et al. A 15 MIPS 32b microprocessor. *IEEE International Solid-State Circuits Conference*, ISSCC 87:26–27, 1987.

[10] Takaaki Komatsu et al. A 35-ns 1-Mb SRAM. *IEEE International Solid-State Circuits Conference*, ISSCC 87:258–259, 1987.

[11] Otto G. Folberth. Hürden und Grenzen bei der Miniaturisierung digitaler Elektronik. *Elektronische Rechenanlagen*, 25(6):45–55, 1983.

[12] Otto G. Folberth and Clemens Hackl, editors. *Der Informationsbegriff in Technik und Wissenschaft*, pages 139–158. Oldenbourg, München, 1986.

[13] Edward Fredkin and Tommaso Toffoli. Conservative logic. *International Journal of Theoretical Physics*, 21(3/4):219–253, 1982.

[14] Ed. Jacques Tiberghien. *New computer Architectures*. Academic Press, London, 1984.

[15] Robert W. Keyes. Miniaturization of electronics and its limits. *IBM Journal of Research and Development*, 32(1):24–28, 1988.

[16] Robert W. Keyes. Physical limits in digital electronics. *Proceedings of the IEEE*, 63(5):740–767, 1975.

[17] Robert W. Keyes. *The Physics of VLSI Systems*. Addison-Wesley, Reading, 1987.

[18] Teuvo Kohonen. *Associative Memory: A System-Theoretical Approach*. Springer, Heidelberg, 1977.

[19] Rolf Landauer. Irreversibility and heat generation in the computing process. *IBM Journal of Research and Development*, 5:183–191, 1961.

[20] Carver Mead and Lynn Conway. *Introduction to VLSI Systems*. Addison-Wesley, Reading, 1980.

[21] Emerson W. Pugh. Technology assessment. *Proceedings of the IEEE*, 73(12), 1985.

[22] Brian Randell. *The Origins of Digital Computers: Selected Papers*. Springer, Heidelberg, 1975.

[23] Charles L. Seitz and Juri Matisoo. Engineering limits on computer performance. *Physics Today*, 740–767, 5 1984.

[24] D.B. Tuckerman and R.F.W Pease. High-performance heat sinking for VLSI. *IEEE Electron Device Letters*, EDL-2(5):126–129, 1981.

PARALLEL COMPUTING

D.J.Wallace

Physics Department, University of Edinburgh, EDINBURGH EH9 3JZ.

1. INTRODUCTION

The aim of these lectures is to provide a basic introduction to the present status of parallel computing, and to some of the key issues in it. In the context of this school I should make it clear that I am talking only about computers which are conventional in the sense that they rely on semiconductor technology. This topic is pertinent to this school on optical computing for several reasons: the development of parallel algorithms and strategies may be of mutual benefit in the two technologies; the richness of architecture in parallel semiconductor systems may be a stimulus to the development of optical ones; parallel computers will spearhead the next surge in performance and thereby set the standard against which optical systems must be assessed; and in any event, parallel computing is of considerable intellectual interest and challenge in its own right.

Some preliminary comments should help to explain the tone of the lectures. It is not practical in the time or space available to enter into technical detail in any depth, nor into a history of the many (un)commercial developments, nor a catalogue of machines on the present market. Rather I

shall aim to provide some perspective, some motivation, expose the key
issues, including performance aspects, and summarise our experience at
Edinburgh in applications of two very different parallel architectures. It
will also be obvious that much of the emphasis is on numerical computing;
this should be seen as a result of my own background and not in any way a
reflection on the significance of parallelism for symbolic computation.

2. A PERSPECTIVE

At a practical level the gain in speed of computers - by a factor of roughly
a million in the past 30 years - is due only in part to increases in the
intrinsic speed of their components, which accounts for a factor of roughly
1000. The other factor of 1000 is due to the implementation of parallelism.
For example on the large scale, input and output from the computer are dealt
with separately from the actual computation, and on a finer scale the
multiplication of each of the digits of one number into another can be done
simultaneously. The kind of parallelism with which we are concerned in this
set of lectures is rather different: it is how you organise many processors
to tackle a big problem cooperatively.

The idea of doing simultaneous calculations with a large number of
computational units is not a new one - in fact it was recognised by Babbage
last century (Hyman 1982), well before the electronic computer was
conceived. Another early and explicit pointer to the potential of parallel
computing is given by Lewis F. Richardson in his book "Weather prediction by
numerical process" (Richardson 1922). From chapter 11/2, we quote:

    ....If the time-step were 3 hours, then 32 individuals could
just compute two points so as to keep pace with the weather, if
we allow nothing for the very great gain in speed which is
invariably noticed when a complicated operation is divided up
into simpler parts, upon which individuals specialize. If the
co-ordinate chequer were 200 km square in plan, there would be
3200 columns on the complete map of the globe. In the tropics

the weather is often foreknown, so that we may say 2000 active
columns. So that 32 x 2000 = 64,000 computers would be needed
to race the weather for the whole globe. That is a staggering
figure. Perhaps in some years' time it may be possible to
report a simplification of the process. But in any case, the
organization indicated is a central forecast-factory for the
whole globe, or for portions extending to boundaries where the
weather is steady, with individual computers specializing on the
separate equations. Let us hope for their sakes that they are
moved on from time to time to new operations.

   After so much hard reasoning, may one play with a
fantasy?....

His fantasy is illustrated in figure 1 (Lannerback 1984), which shows a
uniformly spaced array of computers (who were people, of course, in
Richardson's time), taking boundary data as required from their neighbours,
and "coordinated by an official of higher rank", who "turns a beam of rosy
light upon any region that is running ahead of the rest, and a blue light
upon those who are behindhand." It is remarkable that so many aspects of
parallel computing are recognised in this early work.

3. PARALLEL COMPUTING: WHY DO IT?

The above discussion has already highlighted the potential increase in
performance, but it is worth expanding on a number of aspects.

3.1. New science needs an increase by orders of magnitude

An increase of computing resource by a factor of 2 makes only marginal
impact on any one scientific problem. Consider for example the central
problem of controlling statistical and systematic errors. Statistical errors
reflect the number of independent configurations sampled or events
generated. To halve the statistical error one needs to increase the number
of configurations sampled, and thus the computing resource, by a factor of

4. In many cases, systematic errors vary roughly linearly with the ratio of the grid size to the linear dimension of the system. Thus, if one is to reduce systematic errors by a factor of 2, one needs a system with 2^3 = 8 times more degrees of freedom. Moreover, the number of updates required to generate significant changes in large scale structures must also be increased, by a further factor of roughly 2^z, where z is approximately 2, if the dynamical process is diffusive. This aspect is particularly important in phenomena like turbulent flow, which has significant structures (eddies) on many length scales.

Figure 1. Richardson's scheme for numerical weather prediction using parallel computers, as illustrated by Lannerback (1984).

As a particularly important example in physics, despite continuing progress, we are still very far from dealing computationally, in a fully controlled way, with the interacting many-electron problem which is at the heart of so many phenomena in condensed matter. Incorporating properly the Pauli exclusion principle for many-fermion systems remains a challenging problem for homogeneous materials. For real materials with defects and impurities, the reliable calculation of properties of great fundamental and practical significance, such as embrittlement and catalysis, makes even greater demands.

Finally, there is of course a host of problems which are only worth doing if they can be performed in real time, for which parallel computing may offer the only solution.

3.2. The limits of silicon technology

Whereas science and engineering need an increase in power by orders of magnitude, improvement in intrinsic properties of silicon-based devices appears to be limited by the need to have conductors sufficiently large to support non-ballistic conduction of electrons; device operation speed will continue to increase, but there seems little scope for orders of magnitude. (For further comments see the lectures by Bohle.) Since vector machines now produce an arithmetic result every machine cycle, we cannot expect to see dramatic advances from them - witness how supercomputer companies have moved towards multi-processor machines or multiple vector pipelines.

Of course new technologies, with intrinsically faster electronic switching characteristics, will emerge: gallium arsenide, heterostructures?, high-temperature superconductors??. In any event, new technologies may also support the advantages of parallel computing, and the theme of this summer school is projected to exploit it in a massive way.

3.3. Parallelism in physical systems

It is clear therefore that if physical problems were amenable only to serial or vector computation, the outlook for obtaining significant increases in the necessary computational power would be bleak. However most computationally demanding scientific problems have enormous inherent parallelism. This claim should be well illustrated by the examples of successful applications later in the lectures. Here we limit ourselves to summarising three forms of parallelism which are common in many problems.

* Event parallelism: Many problems boil down to doing a large number of essentially indepedent computations: experimental data analysis, Monte Carlo event generation, much of low level image processing (on the individual pixels of the image), for example. In such cases it is straightforward to distribute the independent tasks to each of the available processors, the machine then operating as a 'task farm'.

* Geometric parallelism: In many problems, one has to simulate the physical behaviour of the system in a region of space. The first approximation typically involves discretising the space in some way, for example into a square mesh or some other kind of two or three (or higher) dimensional grid, as in Richardson's 'forecast-factory'. In order to implement parallelism in such cases, one may simply decompose the region of interest into subregions, with the computation in each of the subregions assigned to one of the processors.

* Algorithmic parallelism: As recognised also by Richardson, each step of computation may be sufficiently complicated that it can be broken down into a number of smaller steps, each of which may be done in parallel. As a simple example, consider matrix-vector multiply. Each row of the matrix must be independently multiplied into the vector to produce an element if the resultant vector. Thus the elements of the new vector can be computed in parallel.

Of course it would be quite wrong to convey the impression that every scientific problem can be mounted efficiently on parallel hardware. What is indisputable however is that a large subset of problems can be, in very natural ways.

3.4. Cost-effectiveness

Parallel systems have some inbuilt advantages in cost effectiveness; they typically do not demand large numbers of special purpose chips, are amenable to standard VLSI design methods, and the replication of components provides the opportunity of economies of scale even in a single machine. However, the assessment of the relative cost-effectiveness of different machines is fraught with difficulties. The basic measure of performance for numerical computation is 'flops', - floating point operations per second. For cost effectiveness, the crude estimate of Mflops/M$ of capital cost fails to take into account subjective and historical factors such as ease of use and availability of software. Even this measure is hard to quantify, since manufacturers' specifications may be misleading, as an unattainable peak performance is usually quoted; actual performance may be 5% (or less) to 50% of this peak for a particular application. Moreover, direct comparisons between mature and newly-announced machines may be misleading in a period of rapid increase in performance characteristics. In multi-processor systems in which communications between processors are based on message passing, the latency to set up a communication can be as crucial as the actual bandwidth in determining real throughput. Finally, the spectrum of performance of the same parallel machine on different problems is much wider than for a conventional architecture. The highest figures can be quoted for special purpose computers if one counts only the cost of raw silicon and the development effort is not included. Such machines are also likely to be less flexible, and less user-friendly unless considerable system software effort is also expended.

3.5. Why not?

At present of course there is a penalty to be paid for all of these advantages: the parallelism must be harnessed to applications. For the most powerful parallel machines today this does involve recoding. From the perspective of our experience at Edinburgh over the past 8 years, it is our clear view that the gain in cost-effectiveness fully justifies the recoding effort for a wide range of problems. In the future, as parallel systems become more highly developed and our experience at using them increases, this recoding effort will certainly decrease; vector machines will presumably then have to become more price competitive.

4. Issues

4.1. Diversity of System Hardware

It is inappropriate to reproduce here a comprehensive classification scheme
for the wide range of existing and announced commercial machines, but a
glossary of the main distinguishing characteristics exposes some of the key
technical issues.

* SIMD v. MIMD: How much autonomy has an individual processing element in
the multi-processor system? This is the primary classification introduced by
Flynn (1972). In Single Instruction-stream Multiple Data-stream (SIMD)
machines, each processor acts on its own local data with the same operation
at each time-step in the execution of the program. In Multiple
Instruction-stream Multiple Data-stream machines (MIMD), each processor
contains its own cpu, so that it can be independently programmed. Clearly,
MIMD is more flexible in principle than SIMD (and is achieved at little cost
in silicon area for the cpu), but the additional freedom may be gained at
the expense of programming complexity and SIMD is the natural parallelism of
a wide class of scientific and engineering problems.

* Vector v. VLIW v. multiprocessor: There are a number of different routes
to high performance. In machines with vector capability, a sequence, or
vector, of identical arithmetic operations is fed through a 'pipe' which
assembles the arithmetic results so that after some initial latency, one
result is achieved per clock cycle; one may regard this as a form of SIMD
parallelism. Very Long Instruction Word (VLIW) machines have a more general
form of pipelining in which several different operations can be set in
motion simultaneously using the different segments of the instruction word.
These architectures are rather widely exploited to obtain high performance
from a single processing unit, which may also form one of the components of
a parallel system, but it is helpful to distinguish them from genuine
multiprocessor systems whose theoretical performance can in principle be
scaled up indefinitely.

* Shared v. distributed memory: Processors may have access to a single

common 'shared' memory; in this case the system must ensure that contentions due to simultaneous access for read or write are resolved. Access to the shared memory may be through a bus or switch. Alternatively, the memory may be distributed among the processors. In this case, the obvious advantage is that the total bandwidth to memory scales with the number of processors; this must be set against the need to provide adequate communications bandwidth for the other processors to access it.

* Hierarchical v. flat memory: All computers have memory hierarchies, in terms of semiconductor RAM, disc and tape storage; in order to achieve peak performance, all data should be stored in the fastest memory. In machines which rely on fast processing units or shared memory, it is also common for the RAM memory itself to be hierarchical, and the achievement of peak performance makes corresponding demands on the application and the programmer (or compiler). The need for a hierarchy in vector and in shared memory machines implies that the 'time thread' of the computation must be carefully structured; this can be contrasted with the need to effect some form of spatial 'domain decomposition' in distributed memory machines, for which however flat or 'homogeneous' memory may be perfectly adequate.

* Coarse v. fine grain: There is no universally agreed definition of coarse-grain or fine-grain parallelism, although they are widely used. At the simplest level they describe the degree of parallelism of the machine hardware; the smaller the number of processors, the coarser the grain, extreme examples being the CRAY XMP48 with 4 and the Connection Machine with 64K. Another distinction can be made in terms of the memory per node - the larger the memory the coarser the grain. The terms coarse and fine grain are also used in a software context, in terms of the time scales on which the parallel processes are synchronised - the shorter the time scales, the finer the grain. None of these features provides a clear distinction, and they are generally correlated; a smaller number of powerful processors would typically each have larger memory, and truly massively parallel systems are most likely to be operated in SIMD mode, i.e. synchronised at each clock cycle.

* Processor connectivity: This question has received considerable

attention, given the need to communicate data amongst processors in a
distributed memory machine. Complete connectivity, in which every processor
is wired to every other one, is impracticable for systems with more than a
few processors, since it requires $N(N-1)/2$ wires for N processors. There are
two main compromises in current commercial systems: a square grid and a
hypercube. In the case of the square grid, the local connectivity may be
enhanced by 'row' and 'column' broadcasts, as in the Distributed Array
Processor. The topology of the Myrias machine is a pyramid of square grids.
The hypercube generalises the square (4) and cube (8), to 16, 32 etc.
processors. Flexible electronic reconfigurability according to application
needs is a feature of transputer arrays. In the longer term, connectivity
may become a less critical issue than it has appeared to be to date, with
the development of more sophisticated routing capability and deeper
understanding of the key factors.

In practice, manufacturer strategies have resulted in four broad classes of
machine organisation to date: coarse-grain distributed memory MIMD,
coarse-grain shared memory MIMD, and fine-grain and coarse-grain distributed
memory SIMD. Specific examples are given in the discussion on current
options in section 3.

4.2. Software Factors

The present state of system and compiler technology is such that in all
supercomputer or novel architecture systems, although to very different
degrees, considerable effort must be made by the programmer if a performance
approaching the machine's peak is to be attained. The dedicated software
effort may be justified on several grounds, either in terms of the ability
to tackle a problem which would otherwise be impossible, or in terms of
overall cost-effectiveness, or in terms of availability of resource or
throughput: any one of these criteria may pertain to industry or academia.
On the other hand, there can be no doubt that portability of existing code
and compatibility with future systems are key factors if the emerging
parallel computing technologies are to continue their volume expansion into
the mass market of existing conventional machines; these factors would be
less important if algorithm performance and implementation on new hardware

Architecture	Construction	Programmability	Performance
(a) Homogeneous Multicomputer	Straightforward construction Simple node may not match (c,d) in performance	Needs user Decomposition but of simplest type software cost	Best Architecture for getting peak performance at modest
(b) Homogeneous shared memory	Network performance crucial. See also node comment in (a) Difficult to build network if node powerful	Allows Compiler Decomposition Easiest of four architectures to program	Limited by Network
(c) Hierarchical Multicomputer	Easiest way to get peak node performance & cost effective machine	Decomposition methodology harder than (a) but exists	Optimal performance possible at more user effort than for (a)
(d) Hierarchical Shared Memory	Construction harder than (c) but allows easier programming	Can be programmed for peak performance as in (c) but allows compiler generated concurrency	Performance can be as good as (c) but this requires comparable programming techniques. Lower but good performance with compiler generated concurrency

Table 1. Comparison of MIMD architectures, after G.C. Fox, (unpublished).

advanced at the same rate as the hardware performance itself, but this is
unfortunately not the case.

Some remarks on these issues are pertinent:

* Languages and compiler technology: All projections suggest that for
scientific and engineering applications and at least on the more mature
architectures, Fortran will continue to dominate, with increasing use of C.
Automatic vectorising compilers will improve in capability, as will tools
such as the BLAS3 (linear algebra) routines for the effective utilisation of

hierarchical memory machines. For parallel machines, the situation is more complex and speculative. One may be confident that more powerful automatic parallelising compilers and load-balancing tools will be developed for shared memory machines, For distributed memory machines the first generations of these tools are now becoming available in the framework of message-passing between processors, at least for some classes of applications, such as the Fortran 'task farm' and problems with manifest 'geometric' parallelism.

* Operating systems: The emergence of UNIX as the system which has to be available on every machine seems to confirm it as a de facto standard.

* The trade-offs: Table 1, due to G.C. Fox (unpublished), provides a summary of his view of the trade-offs between absolute performance, ease of construction and ease of use, for the commercially inportant class of coarse-grained MIMD machines.

5.   RESOURCES AT EDINBURGH

The existence at Edinburgh of dedicated parallel hardware has been a crucial factor in the expansion of the user community (currently more than 100).

Work on applications of parallel computing began in 1980 with G.S.Pawley's use of the ICL Distributed Array Processor (DAP) at Queen Mary College for molecular dynamics studies. The DAP is an SIMD machine which was comprised, in these first generation versions, of a 64x64 array of simple processing elements, each with 4 or 16 Kbits of associated memory; an ICL 2900 mainframe acted as the host. Each processing element communicates with its four nearest neighbours in the array (North, South, East and West), and the user can arrange the array to have either fixed or cyclic boundary conditions. Each processor can perform only bit-serial arithmetic, that is, operating on one bit at a time, so arithmetic operations must be achieved in software. This means that the DAP offers a flexibility in word length unavailable in conventional computers, which usually allow only 16-bit, 32-bit or 64-bit operations. The DAP is programmed in DAP-FORTRAN; the host machine uses standard FORTRAN. Existing software routines therefore need to be converted to DAP-FORTRAN. Our experience is that it is easy to achieve a performance of around 20Mflops on many problems, with the DAP. For problems with short words, and in particular for parallel bit-manipulation, the DAP is an extremely powerful architecture - see remarks in the examples. For further information on the DAP and its software, see for example (Hockney and Jesshope 1981).

Our early experience of the performance of the machine at Queen Mary College was sufficiently convincing that we were led to prepare a proposal to SERC to site a DAP at Edinburgh. The success of this proposal and the work emerging from it resulted in the gift of a second DAP from ICL and these two machines provided a superb resource, with more than 180 publications emerging from this work, covering a much wider range of science than was envisaged in the original proposal; a brief summary and references are contained in (Bowler et al 1987a). The DAPs were necessarily decommissioned with the replacement of the ICL hosts with new University mainframes at the end of July 1987. Work on this architecture is continuing with the installation of an Active Memory Technology DAP 510 system under an Alvey grant. This 32x32 array is hosted by a SUN Workstation (or microVAX). The minimum memory size is 4 Mbytes, with possible expansion up to 128 Mbytes, removing the major limitation of the mainframe DAPs, which had only 2 Mbytes. A second AMT DAP has recently been installed, principally for protein and DNA sequence analysis.

The range of these results and applications should give some encouragement to the development of parallel optical machines, since in the machine architecture is probably closest in concept to what one might try to achieve with parallel digital optical circuits.

In anticipation of the likely loss of the DAPs, we were fortunate to acquire, in April 1986, with support from the Department of Trade and Industry and the Computer Board, the first Meiko Computing Surface delivered to a University. The Meiko machine (Bowler et al 1987b) is an electronically reconfigurable distributed memory MIMD machine with advanced graphics capability. The processor at each node is the Inmos transputer, consisting

of a 10 MIPS cpu, some on chip RAM, and four communications channels, each
capable of 20 Mbits bidirectional. The logic for communication is integrated
onto the chip so that they can be wired together directly or via a switch as
in the Computing Surface and other machines; the low latency of a few
microseconds to set up comunications is also important. All of these
features can function concurrently. In addition to this functionality of
the T414 chip, the floating point transputer (the T800) also has a 64 bit
floating point unit integrated onto it which is capable of sustaining over 1
Mflops.

The demonstrator machine consisted of 40 T414 transputers each with 256
Kbytes of RAM, and display system, and was hosted by a microVAX. The
impending loss of the DAPs, the hardware reliability and software
environment of the demonstrator system, and an evaluation exercise carried
out over the Summer of 1986, led to the preparation of the Edinburgh
Concurrent Supercomputer proposal in September 1986. Phase One support for
this project has been awarded by the DTI, Computer Board and SERC, providing
a multi-user facility with 32 domains, four display systems and a frame
grabber for real-time video i/o, 6 Gbytes of disc capacity, and compute
resource consisting of 200 floating point transputers (T800s) each with 4
Mbytes of local memory. The multi-user sustem, MMVCS, will shortly be
superseded by MEiKOS, a UNIX-based system which is currently under test. The
facility is run for University and national users by the Edinburgh
University Computing Service. A condition of DTI support was the
establishment of an industrial affiliation scheme, which has already met the
initial target of £500K commitment (in cash and kind). The system will be
expanded early in 1989 with a further 200 T800s and 800 Mbytes of memory, as
well as additional disc capacity. Further general information can be
obtained from the Project Newsletters (Wilson, 1987, Stroud and Wilson
1988), and the Project Directory (Wexler and Wilson, 1988) summarises some
60 projects under way on the machine.

It is anticipated that these resources will continue to be used primarily
for scientific and engineering applications, although Artificial
Intelligence projects are now beginning to emerge also: the Parallel
Architectures Laboratory in the Artificial Intelligence Applications

Institute at Edinburgh has parallel hardware specifically for AI
applications. It is also noteworthy that Bolt Berenek and Newman have
installed a 32-processor Butterfly GP 1000 at their European headquarters on
the Heriot Watt Science Park here.

6. EXAMPLES

The following paragraphs review a selection of successful applications of
the work to date at Edinburgh. The examples are far from exhaustive, and are
slanted somewhat towards the DAP in viewof its potential architectural
interest.

6.1. Experimental data analysis

Most of our examples are based on theoretical modelling and simulation, but
computational demands in data analysis are also becoming increasingly
onerous, so it is pertinent to consider an example in this area. The
potential of "event parallelism" in the analysis of high energy physics data
has already been noted by Glendinning and Hey (1987). The particular example
in condensed matter physics on which we focus here is the calculation of
full resolution corrections in neutron scattering data (Mitchell and Dove
1985).

Typically, a neutron inelastic scattering experiment is designed to measure
the scattering function $S(Q,W)$, which contains information about the
microscopic static and dynamic properties of the system under study.
However, what is actually measured is a convolution of $S(Q,W)$ with some
experimental resolution function which in the general case also depends upon
the four variables Q and W. Such corrections must be made in order to
compare experimental results with theoretical predictions, and can be
computationally intensive.

To ease this problem, a package was written for the ICL DAP which exploits
algorithmic parallelism for each data point, to perform these corrections
for a user-defined model and to fit the resulting theoretical scattering

function to the experimental data. The loop over all data points is in fact
done serially, although it offers further obvious scope for parallelism.
Applications studying scattering by spin waves and phonons in a number of
magnetic materials are reported in (Mitchell and Dove 1985). The DAP program
enables interactive fits to be made in roughly one minute or less, compared
with roughly 20 minutes of CPU time for the equivalent code on a mainframe.
This package was in routine use at Edinburgh, until the decommissioning of
the DAPs; its advantage for enhancing the effectiveness of research work in
this area is obvious.

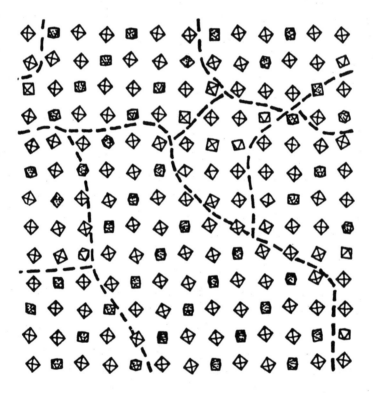

Figure 2.   Two-dimensional section through a simulation (Pawley and Thomas
            1982) SF6 at 25K. The molecules are represented by octahedra.
            there is clear evidence of microdomains of the low-temperature
            phase which involves orientational reordering of the SF6
            molecule.

6.2. Molecular dynamics

This is another area which is manifestly amenable to parallel computation, because the time-stepping is naturally done by the simultaneous calculation of the forces on all the N molecules. Usually the number of processors available is less than or equal to N, but even when they are greater than N they can still be used efficiently either by distributing the force calculation for each molecule across a number of processors, or by running independent simulations in parallel to accumulate statistics. Since realistic modelling of solid-state phase transitions requires a sample large enough that the nucleating phase is not restricted or determined by the imposed boundary conditions, significant simulations offer scope for massive parallelism.

The particular example on which we focus is an early study (Pawley and Thomas 1982) of the plastic-to-crystalline phase transition in sulphur hexafluoride, SF6 , using the DAP. The octahedral shape of the molecule is conducive to the formation of a plastic phase at intermediate temperatures, in which the molecules form, on average, a body-centered cubic lattice, and perform occasional reorientational jumps. When a sample (roughly 13x13x13 bcc unit cells in practice) is simulated at low temperature (25K), however, the temperature of the sample slowly drifts up as equilibration proceeds, requiring extraction of kinetic energy to maintain the nominated temperature. The total potential energy of the system is accordingly falling (and the volume of the constant-pressure sample is decreasing), pointing to a gradual ordering process.

The obvious question is: what is the nature of the new ordered state? Figure 2 is taken from (Pawley and Thomas 1982) ; it represents a two-dimensional section through a sample after such a simulation and shows clearly a mosaic containing microcrystals in which the now triclinic unit cell contains one molecule of one orientation and two molecules of a second orientation. The new structure is well supported by experimental results from neutron powder diffraction refinements (Garg 1977, Dove et al 1987). The figure underlines the necessity of a large enough simulation to support a mosaic of the new structure; clearly simulations with more complicated molecules or ordering require correspondingly even more computational resource.

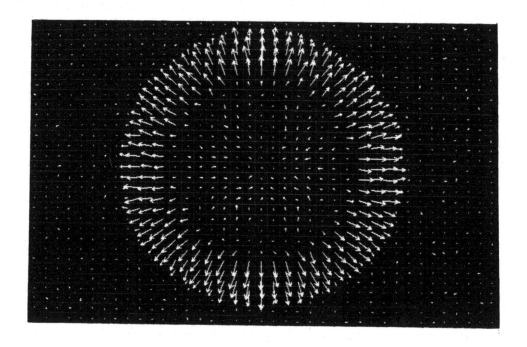

Figure 3. Cellular automaton simulation on the Computing Surface
           (Kenway, McComb and Wylie, unpublished) illustrating how the
           underlying triangular lattice supports a circular wavefront.

6.3. Cellular automata

Cellular automata are arrays of discrete cells which can contain degrees of
freedom which take on discrete values (boolean variables 0 or 1 in the
simplest case). These variables evolve in time according to some transition
rules which are dependent on the state of variables in neighbouring cells

and may be deterministic or stochastic (i.e. affected by noise). In one sense therefore, they can be thought of as primitive molecular dynamics. The motivation for studying them is (at least) two-fold. From the point of view of physics, although their microscopic behaviour may not correspond to any specific physical system, their behaviour on distance scales large compared to the cell size can describe macroscopic continuum phenomena. The spirit here is very much akin to universality phenomena at phase transitions underpinned by renormalisation group theory (for an early review see (Wilson and Kogut 1974)). From the point of view of computation, they are suitable for digital computer simulation, particularly on SIMD computers with powerful parallel bit-manipulation capability like the DAP, Connection Machine or Goodyear MPP.

Microcanonical simulations of the Ising model of a uniaxial ferromagnet can be formulated in this way for example, but the particular case study on which we focus is cellular automaton modelling for fluid flow (Hardy et al 1976, Frisch et al 1986, Salem and Wolfram 1986). These models represent an extension of the lattice gas concept from statistical mechanics to hydrodynamics, with "particles" hopping along bonds of the lattice and scattering from each other according to simple local rules. The primary aim is to ensure that the Navier-Stokes equation emerges on the large scale. It turns out that the discrete symmetry of a square-lattice automaton survives in the macroscopic limit. However, a hexagonal lattice has sufficient symmetry to ensure isotropy, which can also be ensured in three dimensions by allowing hopping beyond nearest neighbours (Frisch et al 1986, Wolfram 1986, Frisch et al 1987).

In figure 3 we show a result obtained by B.J.N. Wylie, using the Edinburgh Computing Surface (Kenway, McComb and Wylie, unpublished). The simulation depicts an expanding circular wavefront. The system permits the specification of barriers of any shape in the flow, and accommodates the interaction of complex flows as in the figure. This type of bit-serial simulation is more ideally suited to the architecture of the DAP than that of the Computing Surface, but the latter's graphical capability was crucial in 1986 in motivating the work and in visualising the results. Whether this approach to turbulent simulation captures hydrodynamic flow effectively, and

will emerge with significant advantages over conventional methods using the
Navier Stokes equations, remain matters for research and debate; the
existence of fine-grain parallel computers has certainly been a major factor
in stimulating that debate.

6.4. Monte Carlo simulation

Several distinct kinds of computation come under this heading. For example,
it is an integral part of radiation metrology generally, and of the
calibration of detector performance in high energy physics experiments in
particular, by studying the response of the detector to events generated at
random according to some model of the collision processes. In contrast, the
kind of calculation on which we focus is the Monte Carlo simulation of the
canonical ensemble in some thermodynamic problem. In particular, we consider
an example from the study of critical phenomena at phase transitions. Such
problems are very demanding in computing resources, because the critical
singularities emerge only in the infinite volume limit, and their true
universal values may be obscured by the finite-size effects which are always
present in the finite systems studied on a computer. Moreover, whereas
configurations may evolve rapidly on the small (e.g. lattice) scale under
the simulation, on the scale of the correlation length, ξ, they evolve
with a characteristic time which increases as ξ^z where $z \sim 2.0$.
Thus, to obtain reliable results we need large systems and long runs. The
calculations are well suited for parallel computation, because any subset of
variables can be updated simultaneously and independently, provided there is
no direct interaction between them in the Hamiltonian.

The particular study we report here concerns the issue of hyperscaling in
the three-dimensional Ising model (Wall 1986, Freedman and Baker 1982). The
question is whether in this model the relationships between the critical
exponents governing thermal properties and those governing the correlation
length are as predicted by the renormalisation group (Wilson and Kogut 1974)
(and hyperscaling arguments, see (Fisher 1983) and references therein).
Freedman and Baker (1982) studied this problem by considering the quantity

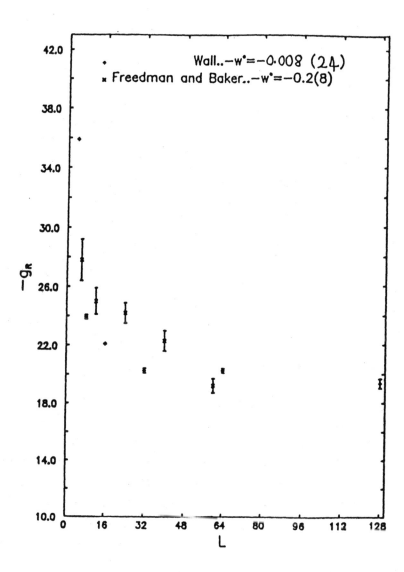

Figure 4. Plot of the renormalised coupling constant g_R of the
three-dimensional Ising model obtained (x) by Freedman and
Baker (1982), and (+) by Wall (1986), plotted as a function of
the linear dimension L of the lattice; hyperscaling is valid
if g_R tends to a constant asymptotically.

$$g_R = \frac{\chi^{(4)}}{(\chi^{(2)})^2 \, \xi^d}$$

where $\chi^{(2)}$ is the two-point magnetisation cumulant (the susceptibility), $\chi^{(4)}$ is the four-point cumulant, and ξ is the correlation length.

If one varies the system size L while keeping the ratio ξ/L constant, general arguments indicate that (for large enough L), g_R should behave like

$$g_R \propto L^{-\omega^*}$$

where ω^* is a critical exponent; according to hyperscaling and the renormalisation group, for the 3-dimensional Ising model, ω^* should be zero and γ_R should tend to a non-zero constant. Numerical simulation by Freedman and Baker (1982) suggested the value $\omega^* = 0.20(8)$. This calculation was extended on the ICL DAP (Wall 1986), with a lattice of up to 128^3 spins, generating (in the case of the largest lattice) some 70 million configurations using code running at more than 200 million single spin update attempts per second. The combined results are shown in figure 4. The levelling off of the cumulant ratio is rather convincing in the high statistics DAP results, and a fit to the data yielded $\omega^* = 0.008(24)$, in good agreement with the renormalisation group predictions.

6.5. Percolation

It is well known that percolation processes provide examples of critical phenomena; their reliable study can therefore be as computationally demanding as phase transitions.

In its simplest form, the problem concerns the statistical properties of

clusters formed by depositing sites (or bonds) on a lattice with some
probability, p. In particular, one is concerned with the universal critical
exponents governing the scale of the clusters, their total number, size
distribution, etc., in the neighbourhood of the critical concentration pc –
the smallest value of p necessary to generate an infinite cluster. The
conventional theoretical analysis predicts that, for example, the singular
part of the mean number of clusters per site has the form (Stauffer 1979)

$$K_s(p) = D \ |p-p_c|^{2-\alpha}$$

for p near p_c, where $\alpha=-\frac{2}{3}$. However, it has been suggested recently on the
basis of numerical work and series expansion studies (Jug 1985) that the
singular part K_s should have the form

$$K_s(p) = D \ (p-p_c)^2 \ \ln|\ln|p-p_c||$$

This controversy motivated a study of the problem on the ICL DAP (Dewar and
Harris 1987). As part of that work a parallel algorithm was developed for
counting clusters in two dimensions by collapsing them to a single site (or
spanning loop) by using fast parallel bit manipulations on the DAP. Towards
the end of this computation, when most of the clusters have already been
collapsed, the remaining 'active' cluster sites are rather sparsely
distributed across the processor array, and efficiency is correspondingly
low. However, the DAP code (Dewar and Harris 1987) ran some 20 times faster
than a serial algorithm on the CRAY 1 (Jug 1985). This is an interesting
case in which the power of parallel bit-manipulation on the DAP compensates
for its rigid SIMD parallelism.

In exposing the singularity in K, it is convenient to eliminate the leading
regular terms by considering the third derivative with respect to p, K'''.
The results for this quantity at the critical concentration, as a function
of lattice size L, are shown in figure 5. The DAP results are in complete
agreement with the conventional scaling result: $K''' \propto L^{\frac{1}{4}}$.

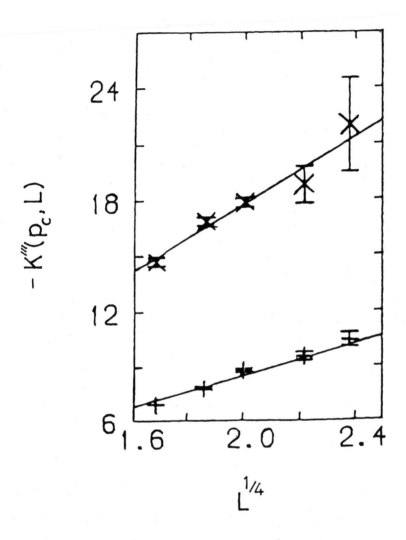

Figure 5. In the conventional theory of percolation, the third
derivative of the cluster density is predicted to diverge at
the percolation concentration as $L^{1/4}$ in two dimensions, where L
is the linear dimension of the system. These results are taken
from a simulation on the ICL DAP using a fast parallel
algorithm for cluster counting (Dewar and Harris 1987).

6.6. Neural Network Models

The remarkable processing capabilities of the nervous system - vision, speech-recognition, motor control, reasoning, association etc - are achieved despite the fact that the typical timescales of the biological 'wet-ware' are of the order of milliseconds, rather than the nanoseconds of modern silicon computers. Neural network models attempt to capture the key ingredients responsible for these faculties; among them is undoubtedly massive parallelism.

A wide range of models has been developed for a variety of purposes; they differ in structure and details, but share some common features:

 - they contain nodes, or units, which are extreme
   simplifications of neurons, in the sense that the state of
   each node is usually described by a single real variable,
   representing its firing activity;

 - the nodes are connected, usually in pairs, so that the state
   of one node affects the potential of all the nodes to which it
   is connected according to the weight, or strength, of the
   connection;

 - the new state of a node is a non-linear function of the
   potential created by the firing activity of the other neurons;

 - input to the network is done by setting the states of a subset
   of the nodes to specified values; this sets up an image or
   pattern of activity on these 'input' nodes;

 - the processing takes place through the evolution of the states
   of all the nodes on the net, according to the details of the
   dynamics and the particular connection strengths, until some
   output activity can be read from another, possibly different,
   set of nodes;

- the training of the net is the process whereby the values of
  the connection strengths are modified in order to achieve the
  desired processing for a set of training data.

For example, for image processing, one can imagine that the array of pixel
values from the input image is mapped on to the states of the array of input
nodes. The image processing is effected by the dynamics of the net
producing the enhanced image or the features identified in the image etc, at
the output nodes. The training data would consist of a set of possibly noisy
images with their known, desired output. For an application in medical
diagnosis, the input data would be an encoding of the symptoms, and the
target output would be the diagnosis, and possibly recommended treatment.
Models based on these ideas are not new. Much of the current effort can be
traced to the seminal work of McCulloch and Pitts (1943), and Hebb (1949).
In the 1960s, Rosenblatt (1962) developed the 'perceptron' model for pattern
classification, and Widrow and collaborators (1960) demonstrated weather
prediction for southern California using the 'adaline' analogue computer
based on neural net principles. An excellent analysis and critique of the
perceptron theory is given by Minsky and Papert (1969). The recent
resurgence of interest was in large measure stimulated by the development of
networks which go beyond the limitations of the perceptron model; reviews
can be found in (Hinton and Anderson 1981, Rumelhart et al 1986, Denker
1986, Grossberg 1987). The potential of optical computing to overcome the
connectivity demands in many models and applications has also resulted in
considerable interest; see the lectures by Psaltis.

6.6.1. Optimisation by analogue neurons

         Analogue neurons were introduced by Hopfield and Tank (1985) as a
general technique for optimisation problems involving boolean variables. The
method was applied by them originally to the travelling salesman problem
(see next section), although it turns out to be not very effective at that
particular task (Wilson and Pawley 1987). It has also been used to perform
analogue to digital conversion (Tank and Hopfield 1985) and load balancing
in parallel computing (Fox and Furmanski 1987). We focus here on image

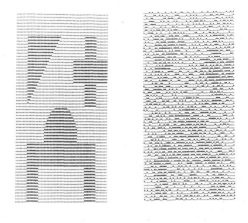

Figure 6.   Image restoration by analogue neurons (Forrest 1987), in
            the framework of Geman and Geman (1984). The top two images
            are an original, and with 30% noise. The three restored images
            at the bottom were produced using analogue neurons, gradient
            descent, and a majority rule, from left to right respectively.

restoration, using the algorithm of Geman and Geman (1984) for binary images
which have been corrupted by noise, which had previously been studied by
Murray et al (1986) using the simulated annealing algorithm (Kirkpatrick et
al 1983). The analogue neuron scheme (Forrest 1987) involves representing
the array of pixel intensities as a network of neurons, each of which can

'fire' on a continuous scale from non-firing ('black' pixel) to fully-firing ('white' pixel). The dynamics of the pixel interactions ('neural activities') is controlled by a cost function determined by the input data and by a priori assumptions about the statistical properties of the 'clean' images (for example, "edges are rare"). The 'best' restored image is that which minimises this cost function, and the dynamics is designed to achieve at least a good approximation to this. This problem is intrinsically SIMD-parallel, with short-range communications, so that it can be expected to run efficiently on most parallel machines; at Edinburgh it was studied both on the DAP (Forrest 1987) and on the Computing Surface (D.Roweth, unpublished). The performance of the restoration by this analogue neural network method was compared to that of a simple majority-rule scheme (where each neuron continually adopts the intensity of the majority of its four nearest neighbours - or remains unchanged if exactly two of its neighbours are 'white' - until the image stabilises). A third restoration method - performing a gradient descent - was achieved by restricting the neuron firing rates to discrete values ('on'/'off'). The analogue neural network method consistently finds lower cost solutions (Forrest 1987) than these schemes, as illustrated in figure 6.

6.6.2. The elastic net

In a recent paper, Durbin and Willshaw (1987) described what they called the 'elastic net' method for solving the travelling salesman problem in the plane. The method is generally applicable to problems involving mapping between spaces of different dimensions (Mitchison and Durbin 1986). For the travelling salesman problem, a closed loop of elastic 'string' is placed in the plane containing the cities which the salesman is to visit, and then slowly deformed into a path which connects all the cities. The elastic string is modelled as a set of discrete points, each of which is connected to its two neighbours. Attractive forces between each point and its neighbours hold the string together, while each city exerts an attractive force on each point, pulling the point towards it. As the point-to-point forces are relaxed and the city-to-point forces strengthened, the string is gradually deformed into a path. The string dynamics is intrinsically parallel, and has been implemented by a straightforward

'domain decomposition' on the Computing Surface (D.Roweth and G.V.Wilson,
unpublished); for a comparative evaluation of the DAP and Computing Surface
for this problem see (Simmen and Wilson, 1988). The performance is
illustrated in figure 7 from (Durbin and Willshaw 1987). The parallel
elastic net solution to the travelling salesman problem is another example
which becomes more efficient as it becomes larger. Since the number of
points which must be used to model the elastic string grows linearly with
the number of cities, the amount of calculation at each step grows as
N^2. However, the amount of communication only grows as N, so the ratio
of calculation to communication improves as the problems being solved grow
larger.

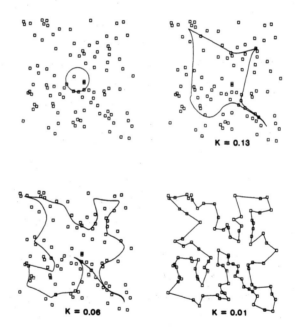

Figure 7. Example from (Durbin and Willshaw 1987) of the progress of the
          elastic net method for 100 cities randomly distributed on the
          unit square. The initial path is a circle, which evolves over
          the next three pictures to a low-cost, legal tour of the
          cities.

6.7. Protein sequence analysis

Molecular Biology has been revolutionised by the development of fast
sequencing techniques for nucleic acids. The rate of acquisition of protein
sequence data has correspondingly accelerated, and this has led to the
urgent need for adequate comparative sequence analysis, to promote the
efficient use of other research resources. Proteins and nucleic acids (the
genetic material) are linear polymers whose sequences may be represented by
character strings, with a 20-letter alphabet for proteins and a 4-letter
alphabet for nucleic acids. The international database collections of
sequences are prime resources for molecular biological research. These
databases are currently small; the protein database has approximately one
million characters of sequence information, and the genetic base has ten
million, but already the task of searching them has led to the development
of a number of approximate methods for making comparisons. However, the
application of the exhaustive inexact string-matching algorithms reviewed by
Sellers (1980) has been beyond the capacity of many workstations and
mainframe computers. The situation will deteriorate further, as the
databases are growing exponentially, doubling in size every two years or
less. A suite of programs for exhaustive inexact string-matching has been
developed for the DAP by Lyall et al (1986); the most valuable of these,
especially for the case of novel proteins, has implemented the 'Best Local
Similarity' algorithm of Smith and Waterman (1981). The programs exploit the
variable wordlength flexibility on the DAP, and perform 4096 comparisons
simultaneously (Coulson et al 1987). They have been extensively applied in
several thousand searches, leading to discoveries of biological
significance. These include the relationship of the cystic fibrosis antigen
to the bovine s-100a alpha protein chain (Dorin et al 1987), and the
relationship of vitellogenins in Drosophila melanogaster to porcine triasyl
glycerol lipase (Bownes et al 1988). A similarity between prokaryotic and
eukaryotic cell cycle proteins has also been discovered (Robinson et al
1987). This work is planned to continue on the AMT DAPs and on the Computing
Surface, where the latter's Fortran Farm facility supports parallel
comparisons transparently across the database. This is one particular
example of the growing application area of parallel databases.

6.8. Medical Imaging

New non-invasive medical techniques such as NMR imaging are being used in
bodyscanners to produce three-dimensional (tomographic) images of the body.
There have been several attempts over the past few years to produce systems
which are capable of manipulating and displaying the vast quantities of data
which are required to generate each image, but a satisfactory performance
has still not been achieved. Three dimensional image processing suffers from
the same speed problems as in two dimensions, but exaggerated by at least
one order of magnitude since there are ten or more 2-D sections in one 3-D
image, and the image-processing operations are inherently more complex. A
key problem in implementing this type of processing on a parallel machine is
to achieve a high efficiency of processor usage while minimising
communications overheads. This is typically straightforward for low-level
image processing, but less so for intermediate and high-level functions. For
example, the common requirement of tracking a surface through the data could
very easily result in one processor at a time doing all the work, with all
the rest lying idle. In particular, the difficulty of mapping a three
dimensional image on to a two dimensional array of processors means that
data at the boundary may need to be passed not to an immediately adjacent
neighbour in the array (as for example in the simulation of fluid flow,
where the same local operations can be performed globally right across the
entire array of processors), but to some more remote processor, depending on
local conditions within the data. While most of the previous examples have
been well suited for SIMD parallelism, the higher level processing in this
example certainly benefits from the additional flexibility offered by MIMD.
The first phase in handling NMR bodyscan data on the Computing Surface is
now complete (Norman 1987). The system includes a generalisation of the
Zucker-Hummel surface detection operator to a non-square metric, a three
dimensional surface display, and an implementation of a novel and completely
asynchronous arbitrary communications network which carries messages around
the processors, ensuring that all are operating as closely as possible to
maximum efficiency. A typical display is shown in figure 8. The computation
runs on 40 T414 transputers at 500 times the speed of a SUN 2, and the
display speed is some 50 times greater.

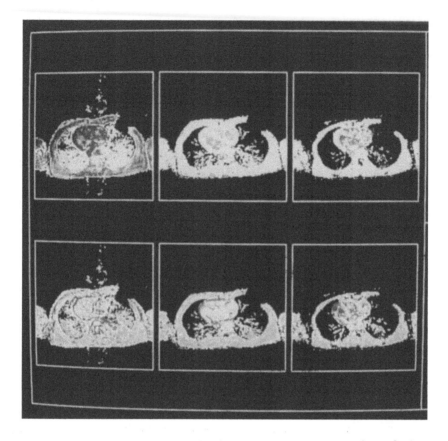

Figure 8. Three NMR datasets showing the same cross-section of the
           human thorax. The top row are smoothed images after
           thresholding. The bottom layer show the results of applying
           the Zucker-Hummel Surface operator to the top row.

7. CONCLUDING REMARKS

This set of lectures has been intended as a guide to some of the issues in
parallel computing, particularly from the viewpoint of a users and
applications. The interested reader is urged to consult for example (Fox et
al, 1988) in addition to the papers already cited, for further details of
recent work and references. The examples of applications and results are

also highly selective, but should serve to underline two major points;
parallel computing is now established as an increasingly important tool, and
apparently rather rigid arrays with bit-serial capabilities at each node can
be remarkably flexible devices.

8. ACKNOWLEDGMENTS

I am grateful to many colleagues at Edinburgh and elsewhere, without whose
efforts the material for this talk would not have existed.

9. REFERENCES

Bowler, K.C., Bruce, A.D., Kenway, R.D., Pawley, G.S. and Wallace, D.J.,
1987a, Scientific Computation on the Edinburgh DAPs, Final Report,
University of Edinburgh Internal Report.

Bowler, K.C., Kenway, R.D., Pawley, G.S. and Roweth, D., 1987b, An
Introduction to Occam 2 Programming, Chartwell-Bratt, Lund.

Bownes, M., Shirras, A., Blair, M., Collins, J. and Coulson, A., 1988,
Proc. Nat. Acad. Sci., 85, in press.

Coulson, A.F.W., Collins, J.F. and Lyall, A., 1987, The Computer
Journal, 30, 420-424.

Denker, J.S., ed, 1986, Neural Networks for Computing, AIP Conference
Proceedings 151, Am. Inst. Phys., New York.

Dewar, R., and Harris, C.K., 1987, J. Phys. A, 20, 985-993.

Dorin, J.R., Novak, M., Hill, R.E., Brock, D.J.H., Secher, D.S. and van
Heyningen, V., 1987, Nature, 326, 614-617.

Dove, M.T., Powell, B.M., Pawley, G.S., and Bartell, L.S., 1987, in preparation.

Durbin, R. and Willshaw, D.J., 1987, Nature, 326, 689-691.

Fisher, M.E., 1983, Critical Phenomena, In Lecture Notes in Physics, (ed Hahne, F.J.W.), 186, Springer-Verlag, Berlin.

Fox, G.C., Johnson, M.A., Lyzenga, G.A., Otto, S.W., Salmon, J.K., and Walker, D. 1988, Solving Scientific Problems on Concurrent Processors. Prentice Hall: New Jersey.

Flynn, M.J., 1972, IEEE Trans. Comput. C21 948.

Forrest, B.M., 1987, Proc. of Parallel Architectures and Computer Vision Workshop, Oxford, in press.

Fox, G.C. and Furmanski, W., 1987, Caltech preprint C3P 363.

Freedman, B.A., and Baker Jr., J., 1982, J. Phys. A, 15, L715-L721.

Frisch, U., Hasslacher, B., and Pomeau, Y., 1986, Phys. Rev. Lett., 56 1505-1508.

Frisch, U., d'Humieres, D., Hasslacher, B., Lallemand, P., Pomeau, Y., and Rivet, J.-P., 1987, Complex Systems, 4, no. 1, in press.

Garg, S.K., 1977, J. Chem. Phys., 66, 2517-2524.

Geman, S. and Geman, D., 1984, IEEE Trans. PAMI, 5, 721-741.

Glendinning, I. and Hey, A.J.G., 1987, Computer Physics Communications, 45, 367-371.

Grossberg, S., ed, 1987, The Adaptive Brain, vols 1 & 2, North Holland.

Hardy, J., de Pazzis, O., and Pomeau, Y., 1976, Phys. Rev. A, 13, 1949-1961.

Hebb, D.O., 1949, The Organisation of Behaviour, Wiley, New York.

Hinton, G.E., and Anderson, J.A., eds, 1981, Parallel Models of Associative Geometry, Lawrence Erlbaum, Hillsdale, New Jersey.

Hockney, R.W. and Jesshope, C.R., 1981, Parallel Computers, Adam Hilger, Bristol.

Hopfield, J.J. and Tank, D.W., 1985, Biological Cybernetics, 52, 141-152.

Hyman, A., 1982, Charles Babbage; Pioneer of the Computer, Oxford University Press (page 242).

Jug, G., 1985, Phys. Rev. Lett., 55, 1343-1346.

Kirkpatrick, S., Gelatt, C.D. and Vecchi, M.P., 1983, Science, 220, 671-680.

Lannerback, A., 1984, Dagens Nyheter, (Sweden), page 35.

Lyall, A., Hill, C., Collins, J.F. and Coulson, A.F.W., 1986, In Parallel Computing '85, (eds Feilmeier, M., Joubert, G. and Schendel, U.), pp. 235-240, Amsterdam: North-Holland.

McCulloch, W.S., and Pitts, W.A., 1943, Bull. Math. Biophys., 5, 115-133.

Minsky, M., and Papert, S., 1969, Perceptrons: An Introduction to Computational Geometry, MIT Press.

Mitchell, P.W., and Dove, M.T., 1985, J. Appl. Cryst., 18, 493-498.

Mitchison, G.J. and Durbin, R., 1986, SIAM J. Alg. Disc. Meth., 7, 571.

Murray, D.W., Kashko, A. and Buxton, H., 1986, IVC, 3, 133-142.

Norman, M.G., 1987, A Three-dimensional Image Processing Program for a Parallel Computer, M.Sc. Thesis, Dept of Artificial Intelligence, University of Edinburgh.

Pawley, G.S., and Thomas, G.W., 1982, Phys. Rev. Lett., 48, 410-413.

Richardson, L.F., 1922, Weather Prediction by Numerical Process, Cambridge University Press, London (republished by Dover Publications, New York, 1965).

Robinson, A.C., Collins, J.F. and Donachie, W.D., 1987, Nature, 328, 766.

Rosenblatt, F., 1962, Principles of Neurodynamics, Spartan Books, New York.

Rumelhart, D.E., McClelland, J.L., and the PDP Research Group, 1986, Parallel Distributed Processing: Explorations in the Micro-structure of Cognition, vols 1 & 2, Bradford Books, Cambridge MA.

Salem, J. and Wolfram, S., 1986, In Theory and Applications of Cellular Automata, (ed Wolfram, S.), pp. 362, Singapore: World Scientific. Sellers, P.H., 1980, J. Algorithms, 1, 359-373.

Smith, T.F. and Waterman, M.S., 1981, J. Molec. Biol., 147, 195-197.

Stauffer, D., 1979, Phys. Reports C, 54, 1-74.

Tank, D.W. and Hopfield, J.J., 1985, AT&T Bell Labs preprint.

Wall, C.E., 1986, Numerical investigation of hyperscaling and real space renormalisation group transformations in the three-dimensional Ising model, Ph.D. thesis, University of Edinburgh.

Wexler, J. and Wilson, G.V., 1988, Edinburgh Concurrent Supercomputer Project Directory.

Widrow, B. and Hoff, M.E., 1960, IRE WESCON Conv Record part 4, 96.

Wilson, G.V. (ed.), 1987, Edinburgh Concurrent Supercomputer Newsletters, 1 2 & 3.

Wilson, G.V. and Pawley, G.S., 1987, Biological Cybernetics, in press.

Wilson, K.G. and Kogut, J., 1974, Phys. Lett., 12C, 75-200.

Wolfram, S., 1986, J. Stat. Phys., 45 471-526.

9 780905 945170